Pharaoh's Flowers

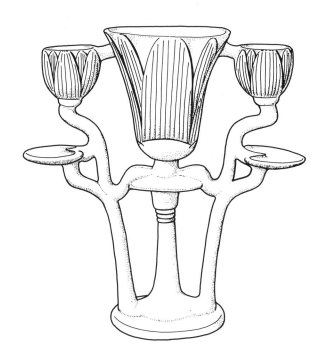

Royal Botanic Gardens, Kew

Pharaoh's Flowers

The Botanical Treasures of Tutankhamun

F Nigel Hepper

London: HMSO

F Nigel Hepper was formerly head of the Tropical African Section and Assistant Keeper of the Herbarium at Kew. He has written widely on archaeobotany, and is the author of *Planting a Bible Garden*, and many publications on tropical flora.

© The Board of Trustees of the Royal Botanic Gardens, Kew, 1990
First published 1990

British Library Cataloguing in Publication Data

A CIP catalogue record for this book is available from the British Library

ISBN 0 11 250040 4

Front cover: a carved ivory box-lid showing Pharaoh Tutankhamun and Queen Ankhesenamun in a garden, with splendid floral bouquets behind them.
Photo: Robert Harding Picture Library

Back cover: the solid gold mask of Tutankhamun, which covered the mummy's head. The wide collar is decorated with lotus waterlily petals in semiprecious stone.
Photo: F N Hepper

Half-title page: an alabaster lamp in the form of a lotus waterlily. From E Wilson, *Ancient Egyptian Designs*, London: British Museum, 1986.

Frontispiece: a detail from the cedar wood chair back shown on p 39.

Title page: an elaborate floral bouquet design, from the furniture found in Tutankhamun's tomb. From E Wilson, *Ancient Egyptian Designs*

pp viii, 7 and 49: lily motifs. From E Wilson, *Ancient Egyptian Designs*

HMSO publications are available from:

HMSO Publications Centre
(Mail and telephone orders only)
PO Box 276, London, SW8 5DT
Telephone orders 071-873 9090
General enquiries 071-873 0011
(queuing system in operation for both numbers)

HMSO bookshops
49 High Holborn, London, WC1V 6HB
071-873 0011 (counter service only)
258 Broad Street, Birmingham, B1 2HE
021-643 3740
Southey House, 33 Wine Street, Bristol, BS1 2BQ
(0272) 264306
9–21 Princess Street, Manchester, M60 8AS
061-834 7201
80 Chichester Street, Belfast, BT1 4JY
(0232) 238451
71 Lothian Road, Edinburgh, EH3 9AZ
031-228 4181

HMSO's Accredited Agents
(see Yellow Pages)

and through good booksellers

Printed in the United Kingdom for HMSO
Dd 291759   C60   9/90

Contents

Preface

Ancient Egypt has a particular fascination, even for people who know little about its civilisation, and tales of golden treasure and weird curses only add to its mystique. Visitors to almost any of the world's major museums can see Egyptian objects and statues inscribed with hieroglyphs or picture writing. Even the owners of these articles may be seen as they were mummified and buried thousands of years ago along with their worldly treasures in pyramids and rock-cut tombs.

Such tombs were always likely to be robbed of their treasures, so elaborate devices were made to foil thieves. Just a few graves have reached the present time intact, but most have been ransacked for valuables, leaving behind the seeds and baskets, linen and papyrus, timber and resins, that were of no value to the thieves. Egyptologists took a long time to appreciate their significance – archaeologists were more interested in the pots than their contents – although they were the reason for the pot being left there in the first place! Even Tutankhamun's tomb was not immune to theft, but fortunately the bulk of the objects were left in place.

Tutankhamun was buried with a reed wand which, according to the inscription on it, 'was cut with his Majesty's own hand'. His body was garlanded with fresh flowers which, over 3,000 years later, are still recognisable. The young king's gilded furniture was buried with him, together with his childhood ebony chair and linen clothes, bark-encrusted bows and reed arrows, perfumes from exotic plants and a host of other items of botanical origin. This book ranges across all of these objects made from plant material. It does not attempt to be comprehensive archaeologically but in botanical terms it looks beyond the flowers to timbers hidden by gold leaf, to dried-up ointment in alabaster jars, and to botanical motifs on chair backs or as lamps. We shall see the food and drink prepared for pharaoh, and even the gaming boards ready for eternal playing.

Readers will notice that there are allusions in the text

Photo: F N Hepper

By Tutankhamun's time the Great Pyramid and Sphinx at Giza were already historic monuments over 1,200 years old.

to relevant passages of the Bible, especially the Old Testament, where Egypt is mentioned. Many more such references could be found, as Egypt played an important role in biblical history and it has had a great influence on the culture of neighbouring nations.

Each chapter consists of two parts. The first section describes the objects found – wreaths, furniture, textiles, etc. – and the second describes individual plant species and the ways in which they were used. Cross-references betwen the two sections are provided throughout. Drawings and photographs of the plants are fully integrated with the text. These show what the species look like as living plants as well as their appearance as dried specimens and motifs in art. The final section of the book, 'Further Reading', will enable archaeologists and botanists to follow up other literature on the subject.

Acknowledgements

I am grateful to all who have given either encouragement or expert knowledge or both: at Kew, Professor Grenville Lucas, David Field, Dr David Cutler, Dr Peter Gasson, Valerie Walley and Sylvia FitzGerald, with photographic support from Andrew McRobb and Media Services; the production staff at HMSO; special thanks to Dr David Dixon of University College, London, an Egyptologist with an interest in plants who kindly read the draft with a very keen eye, but any faults remain mine. Dr Renate Germer's account of Tutankhamun's plant material was published before I had completed my draft and it was a valuable source of reference. I thank Dr J Malék and Miss Fiona Strachan, Griffith Institute, Oxford, for permission to reproduce Harry Burton's photographs taken at the tomb with Howard Carter; Dr Nicholas Reeves, Department of Egyptian Antiquities, British Museum; Lord Carnarvon and A W Saxton of Highclere for approving the biographical notes on the fifth Earl; Lady Eva Wilson for permission to use several of the splendid drawings in her book *Ancient Egyptian Designs* (London: British Museum, 1986); and to Professor N El Hadidi, University of Cairo. I am also grateful to: the Robert Harding Picture Library, the *Journal of Egyptian Archaeology*, and also Professor E W Beals (University of Wisconsin), Juliet Pannett and S Carter, for providing additional illustrative material.

This book is dedicated to the memory of
L A BOODLE
of the Jodrell Laboratory,
Royal Botanic Gardens, Kew

Chronological Chart of Ancient Egypt

Date	Egypt	Other Civilisations
Pre-dynastic dates are very uncertain		
3100 BC	Union of Upper (South) and Lower (North) Egypt	
OLD KINGDOM		
2686–2613 BC	3rd Dynasty (at Memphis)	*Troy settled*
2613–2494 BC	4th Dynasty (Great Pyramids)	
2494–2345 BC	5th Dynasty	
2345–2181 BC	6th Dynasty	*Middle Minoan Age in Crete*
First Intermediate Period		
2181–2010 BC	7th–10th Dynasties	
MIDDLE KINGDOM		
2106–1963 BC	11th Dynasty (at Thebes)	*Abraham's journey from Ur to Egypt*
1963–1786 BC	12th Dynasty	
1786–1648 BC	13th Dynasty	*Reign of King Hammurabi in Babylon*
Second Intermediate Period		
1648–1550 BC	14th–17th Dynasties (rule of Hyksos)	*Joseph in Egypt* *Late Minoan Age in Crete*

NEW KINGDOM

1550–1295 BC	18th Dynasty	
1550 BC	**Ahmose I**	
1525 BC	**Amenophis** (Amenhotep) I	
1504 BC	**Tuthmosis I**	
1492 BC	**Tuthmosis II**	
1479(–57) BC	**Queen Hatshepsut**	*Egyptian Empire in Canaan*
1479(–25) BC	**Tuthmosis III**	*Mycenaeans trading with Egypt and Asia*
1427 BC	**Amenophis II**	
1401 BC	**Tuthmosis IV**	
1391 BC	**Amenophis III**	
1353 BC	**Amenophis IV** (Akhenaten) ⎤	*Moses born in Egypt*
1338 BC	**Smenkhkare** (at Amarna)	
1336 BC	**TUTANKHAMUN** (Tutankhaten) ⎦	*Climax of the Hittite Empire in Syria*
1327 BC	**Ay**	
1323 BC	**Horemheb**	
1295–1186 BC	19th Dynasty	*Exodus of the Israelites from Egypt*
1186–1069 BC	20th Dynasty **(Ramessides)**	
1069–525 BC	LATE PERIOD	*David and Solomon in Israel*
525 BC	Cambyses' conquest	
525–332 BC	PERSIAN PERIOD	*The prophet Daniel in Babylon*
332 BC	Alexander's conquest	
332 BC–30 AD	PTOLEMAIC PERIOD	*Jesus Christ taken to Egypt*
30 AD–	ROMAN PERIOD	

Note: many dates are still approximate, especially the earlier ones and those for the Second Intermediate Period. Dates for the First Intermediate Period overlap with the Middle Kingdom. More detailed information is given for the 18th Dynasty than for the others, in order to show the pharaohs who came before and after Tutankhamun. The dates given here are based on papers by KA Kitchen in *High, Middle or Low?*, Gothenburg: Paul Åströms Vorlag, part 1 (1987), and part 2 (1989).

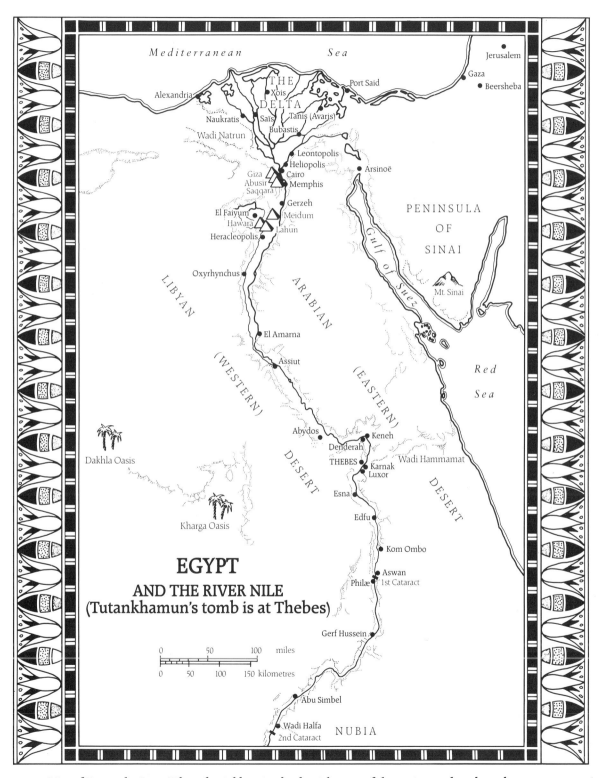

Map of Egypt, the River Nile and neighbouring lands with some of the ancient and modern place-names.

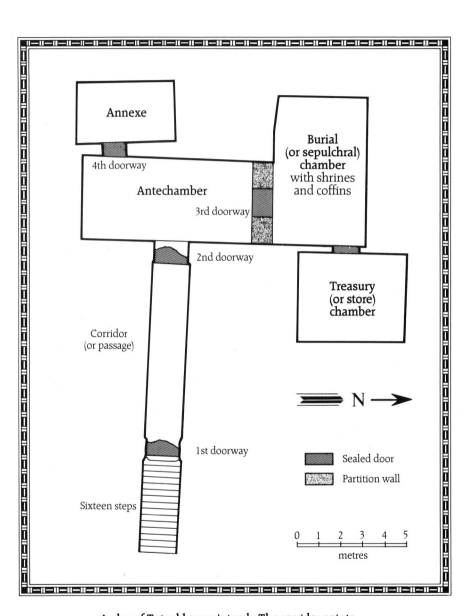

A plan of Tutankhamun's tomb. The corridor points almost exactly west. Based on *The Berkeley Map of the Theban Necropolis*, University of California, 1980.

Introduction

The Life of Tutankhamun

The name of Tutankhamun is known to all of us on account of the fabulous treasures found in his tomb, preserved nearly intact after 3,000 years. We know what he looked like – his face is familiar to us from its image on the golden coffins, and he and his wife Ankhesenamun are depicted on various objects found in the tomb. But who was this king, and when exactly did he reign?

In order to set Tutankhamun in the context of his times, we need to go back a generation or more before his time, to 1353 BC (roughly the middle of the 18th Dynasty), when the reigning pharaoh, Amenophis IV, changed his name to Akhenaten. He also changed the site of government from Thebes to Amarna far to the north; but more importantly he changed the official religion from a plethora of deities to the worship of Aten, the Solar Globe or Disk. This revolution understandably upset the priests at Thebes and threatened them with redundancy. The most powerful deity at Thebes, also threatened by the change, was Amun, although at this time the most widespread cult was that of Osiris, the God-King who was murdered, dismembered and yet put together again. The exact details of the succession of rulers and the timing of their accession and decease in this troubled period are problematic for Egyptologists, but it seems that on the death of Akhenaten and the enigmatic Smenkhares, the throne fell to the ten-year-old Tutankhaten, in 1336 BC.

Before long the young pharaoh, presumably under duress from the priests and people who wanted restoration of Amun worship, changed his name from Tutankhaten to Tutankhamun, transferred his capital from Akhetaten (Amarna) back to Thebes and that huge city was abandoned. It is interesting to note that the gardens of Amarna were a feature of the town and that it was connected by a canal to the Nile. There must have been powerful people behind the throne, in order to make such changes and set about restoring the buildings of Thebes.

A blue lotus waterlily supporting the wooden carved and painted head of the child pharaoh.

One such person was Ay, a senior army officer and adviser, while another was Horemheb, a general with military aspirations. When Tutankhamun died in 1327 BC at the tender age of eighteen it is significant that he was succeeded by both of them in turn! Today, Tutankhamun is notable for what was revealed in his tomb, which was discovered almost intact in 1922, over 3,000 years later. As in life, so in death: the pharaoh was buried with his treasures, for his enjoyment of them in the afterlife, and

1

also with the ingredients necessary to the survival of a wealthy man – bread, wine, fruits, ointments and other materials of plant origin – as we shall see in this book.

The Discovery of Tutankhamun's Treasures

The discovery of Tutankhamun's tomb was no accident. It was the culmination of a prolonged search by a dedicated Egyptologist, Howard Carter, who was sponsored by his patron the 5th Earl of Carnarvon. We can trace this partnership back to 1907, some fifteen years before the great discovery was made. How was it that they joined together as archaeologists?

Lord Carnarvon was a wealthy young man who was an evangelical Christian and quietly generous to needy folk. He inherited his title at the age of twenty-three in 1890, together with the family estate at Highclere Castle in Berkshire. His love of foreign travel was acquired during an exciting sailing cruise around the world after leaving Cambridge; he was an expert shot and was an early motoring enthusiast. Reading became a lifelong interest. Following a car accident in Germany in 1901, when he barely escaped with his life, he found English winters weakened his health, so he visited Egypt in 1903. This triggered his interest in archaeology to such an extent that he was allotted a site at Thebes. As he was a completely untrained 'digger' the French head of the Antiquities Service, Sir Gaston Maspero, decided he could use the Theban site to gain experience, since it had

already been much worked and there was little danger that he would damage it. Indeed after six weeks' effort and expense, Lord Carnarvon's workmen had unearthed only one notable find: a large mummified cat still in its wooden coffin. 'This utter failure,' he later wrote, 'instead of disheartening me had the effect of making me keener than ever.'

Stimulated by the experience, but aware of his limitations, he recruited Howard Carter, who was already a professional Egyptologist. Carter was out of work at the time, having had to leave his post as Inspector-General of Monuments following a dispute between his watchmen at Saqqara and drunken French tourists, and when Lord Carnarvon took him on, at Maspero's suggestion, he was struggling to make ends meet as a self-employed artist at Luxor. Here was a splendid opportunity to excavate in style. From 1907–11 they worked together very productively and in 1912 published a well produced volume *Five years' explorations at Thebes.* In the end they excavated for sixteen years, in the Delta and at Thebes, but the results were disappointingly repetitive until they turned to the royal tombs in the Valley of the Kings opposite Luxor. In 1915 Lord Carnarvon was able to take the concession left by the death of Theodore M Davis in 1914. Davis had excavated in the Valley on behalf of the Metropolitan Museum of Art, New York. In spite of the

Howard Carter. Painting by William Carter, 1924.

Lord Carnarvon reclining.

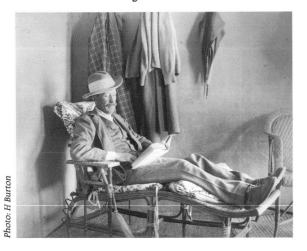

Photo: H Burton

Photo: Griffith Institute

Photo: F N Hepper

The Valley of the Kings in 1963 with the entrance to Tutankhamun's tomb on the right.

First World War, Carter remained in Egypt, continuing to excavate the tomb of Amenophis III, although Lord Carnarvon was in England helping his wife Almina, set up a hospital at Highclere Castle – as well as having surgery himself! His return to Egypt in 1919 proved inopportune owing to civil unrest, yet he was keen to see Carter's excavations of various tombs. From time to time evidence of Tutankhamun's burial was found, presented in the form of lists and accounts on fragments of old pots and limestone chips in the Valley of the Kings, but nobody knew where his tomb was situated. Most of the other pharaohs were known to be buried in the Valley so it was presumed that Tutankhamun's tomb was also there, although Davis had considered it exhausted.

Labourers shifted mountains of debris from likely spots only to have the disappointing sight of bare stone cliffs or workmen's homes. After five years' activity both their patience and their money were nearly exhausted, yet Carter still persisted and he persuaded Lord Carnarvon to try one more season of digging. Despondency turned to elation when three rock-cut steps were exposed. Before long they had dug down fourteen steps to a sealed doorway. Then he cabled Lord Carnarvon, who dropped everything to sail to Egypt, while Carter had the entrance covered up pending his arrival.

The story of the breaking down of the doorway on 22 November 1922, the confirmation that it was indeed the tomb of Tutankhamun and the astonishment at its contents has often been told. A natural impulse to disclose the contents as soon as possible was suppressed in the interests of science and archaeology.

Evidence that the tomb had been entered and resealed appalled these modern discoverers, who feared that the contents might have been taken away already. So it was with apprehension that Carter pierced the wall and gazed in by the light of a candle. When asked whether he could see anything he made the now classic remark 'Yes, I see wonderful things.' This Antechamber was full of a jumble of beautiful pieces of gilded furniture and humble leafy bouquets.

A photographer, Harry Burton of the Metropolitan Museum of Art, New York, happened to be in Egypt and took photographs as the contents were revealed. Many of his original shots are included in this book. A laboratory and photographic dark-room were set up in another tomb nearby and scientific study began. The laboratory was used by A Lucas, a chemist who joined the team and later published results in his book on ancient Egyptian materials (see Further Reading).

The study and clearance were not without problems. The world's press and wealthy sightseers descended on the Valley of the Kings to such an extent that Carter's work ground to a halt. A misunderstanding with the Egyptian government over the terms of the concession was another problem. The sudden death of Lord Carnarvon caused further apprehension, but the work resumed, and culminated in the discovery of the golden sarcophagus and the mummy of Tutankhamun himself. Ever since, popular interest has continued, and much research has been done, resulting in thousands of publications and many films.

The Identification of Tutankhamun's Plant Material

Over the years a wealth of plant material has accumulated from the excavation of the tombs and temples in the Valley of the Kings, and its identification has been undertaken by many botanists of various nationalities. Gradually a picture has been built up of burial customs, local and imported timbers, trade routes and ecological information about the region.

The first scientist to study the living flora of Egypt was a Swede, Petter Forsskål, who perished in 1763 during a royal Danish expedition to Egypt and the Yemen. His results were published in 1775. He was followed by the Frenchman, A D Raffenau-Delile, who was the botanist with Napoleon's expedition and published a superb description of Egypt in 1813. French interest in Egypt has continued ever since – the most notable contribution to the botany of the tombs being that of Victor Loret. The German botanist Georg Schweinfurth (1836–1925) gathered objects of agricultural and botanical interest for the Agricultural Museum which he founded in Cairo, and he sent duplicates to the Berlin-Dahlem botanical museum, where he worked on the material in his later years, publishing may papers on the subject.

However, all this was before Howard Carter's discovery of Tutankhamun's tomb; the botanist who was associated with that event was Professor P E Newberry, OBE, MA.

Percy Newberry was still at school in 1884 when he met Flinders Petrie at the British Museum while unpack-

Professor P E Newberry, who studied the botany of Tutankhamun's tomb when it was discovered. *Journal of Egyptian Archaeology*, vol 36 (1950), pl 12, p 100.

ing his boxes of excavated Ancient Egyptian objects. Petrie even allowed Newberry to draw some of these and the drawings were reproduced in Petrie's book on the ancient Graeco-Egyptian site of Naucratis. By 1888 Newberry had nearly completed his studies at London University, and he had started to identify the plant remains of wreaths and funerary bouquets, as well as fruits and seeds discovered by Petrie in the tombs. At that time this was an unusual subject to study and he was probably the first British Egyptologist to publish on plants.

Later, Percy Newberry became Professor of Egyptology at the University of Liverpool (1906–1919) and Professor of Ancient History and Archaeology, at the University of Cairo (1929–1933). In 1922 he was with Howard Carter during the excavation of Tutankhamun's tomb. Much plant material was discovered there, some of which was identified by Newberry, and is kept at the Cairo Museum, with small samples at Kew. He published an account of these finds in Howard Carter's book (see Further Reading).

Several of the timber samples had been sent to L A Boodle at Kew, who identified them by microscopical examination in the Jodrell Laboratory. He was Assistant Keeper at the time and after his retirement a few years later he was invited to continue the study of Tutankhamun's plant material by Sir Arthur Hill, the Director of Kew, whose help had been sought by Howard Carter in May 1932. Boodle readily agreed to help for a very small

Plate 1 Colourful paintings on the walls of the tomb of Horemheb, the pharaoh who reigned a few years after Tutankhamun's death.

Plate 2 Golden figures of
Tutankhamun on display in Cairo
Museum. Each wears a floral collar and
linen tunic, while the one with a raised
spear stands on a papyrus skiff.

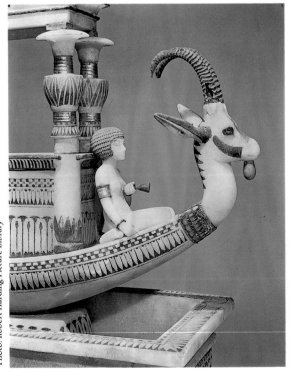

Plate 3 A model boat in alabaster
showing papyrus columns supporting
a canopy. The columns combine the
shapes of the lotus waterlily and
bundles of papyrus sedge.

fee and sent the results to Howard Carter, fully expecting to see the published results of his labours.

Unfortunately the third volume of Carter's book was nearly ready for publication when the Depression, followed by the Second World War, stopped preparation of the definitive works on the tomb. In any case Carter died in March 1939, leaving all his manuscripts to the Griffith Institute at Oxford. His death was followed in August 1941 by that of Boodle, who though a shy and diffident person nevertheless acutely regretted that his study of Tutankhamun's plants never saw the light of day during his lifetime. Boodle was succeeded at the Jodrell Laboratory by Dr C R Metcalfe, who continued research on ancient plant material, including the Tutankhamun finds, in collaboration with Dr L Chalk of Oxford.

No account of the study of the plants of Ancient Egypt can omit Professor Vivi Täckholm (née Laurent). I first met Vivi – as all her friends knew her – as long ago as 1954, soon after I joined Kew Herbarium. Her large jovial figure rapidly engulfed her friends and acquaintances in a great bear hug and she had a toothy smile which soon developed into an infectious laugh. Although Swedish, she lived for most of her life in Cairo. She was born on 7 January 1898 of parents who were both medical doctors.

L A Boodle of the Jodrell Laboratory at Kew, who examined timber, seeds and other plant material from the tomb.

Chalk drawing by Juliet Pannett at Kew

Photo: F N Hepper

The Agricultural Museum, Cairo, where botanical objects from Ancient Egyptian tombs are displayed.

After graduating in botany at the University of Stockholm in 1921, she worked her way around the United States and then returned to Sweden as a journalist. Following her marriage to Professor Gunnar Täckholm in 1926, they both left for the newly founded Faculty of Science at what is now Cairo University, to establish a Botany Department. Gunnar died in 1933 but Vivi decided to continue their work on the flora of Egypt and the creation of a herbarium in Cairo. The first volume of her great *Flora of Egypt*, was published in collaboration with Mohamed Drar in 1941, despite war-time difficulties. Unfortunately the *Flora* was so encyclopaedic in content that the four published volumes contained only the Moncocotyledons and a few of the Dicotyledons. But these four are a marvellous source of reference, for they deal with ancient plant remains as well as the living flora. Her one-volume *Students' Flora of Egypt* (1956, 2nd ed. 1974) is an invaluable tool for any botanist dealing with Egyptian plants.

Vivi's generosity was legendary. She gave books, sweets and scarab beetle seals to students and contacts as prizes and gifts according to the status of the receptor. I was also beneficiary of her hospitality in Cairo in 1963. Her wide knowledge of the history of ancient and modern Egypt made for a fascinating tour of the museums, mosques and churches. Of particular interest to me was the Agricultural Museum and the lesser-known parts of the Cairo Museum where Tutankhamun's treasures are kept. Vivi's interest and knowledge of ancient plant material was extensive, as shown by her splendidly produced popular work *Faraos blomster* ('Pharaoh's Flora') which unfortunately is available only in Swedish. Vivi died suddenly in her eightieth year.

The study of Ancient Egyptian plants goes on, however, both in Egypt and elsewhere. Many of Boodle's identifications were published by A Lucas (in 1926) and all of them by Dr Renate Germer, who is continuing the German Egyptology tradition of Georg Schweinfurth mentioned above. Her booklet, in German, on Tutankhamun's plant material (see references) includes details based on Howard Carter's numbering system (C Nos) for the objects he found, and on Newberry's botanical notes.

Archaeobotanical techniques

Botanical specimens excavated from dry Egyptian tombs are often sufficiently well preserved for identification to be possible without recourse to anatomical examination. This is especially true of fruits, seeds, leaves, flowers and other large objects, but fibres, binding strips, basket material, wood and charcoal must be examined under a microscope.

The aim of microscopical examination is to determine the characteristic cellular structure of the lower epidermis. In order to look at these cells, the upper layers need to be scraped or peeled away. Ancient leaves which have become brittle may need to be simmered in boiling water to soften them, so that this can be done. Hard material needs to be soaked in Jeffrey's solution (10 per

A thin section of oak (*Quercus aegilops*) timber seen under a microscope, showing the distinctive cellular structure and an annual ring.

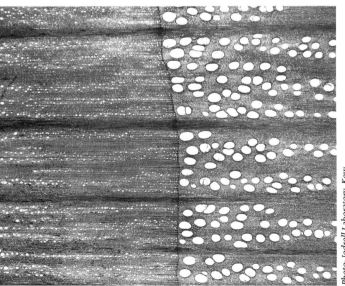

Photo: Jodrell Laboratory, Kew

cent nitric acid, 10 per cent chromic acid) so that it can be sectioned with a razor. First the material is examined under low-power magnification, which reveals the structure of whole cells. In this way it is possible to distinguish at a glance a palm from a grass, a grass from a sedge. High-power magnification can then be used to identify genera and species. The cells become clearer when immersed in Parazone, washed and treated with 70 per cent glycerine. Staining techniques show up the various tissues in different colours. Thus cellulose is coloured green (or blue or yellow) by alcian stain, suffranin stains lignin yellowish, while phenol distinguishes unstainable silica. However, haematoxylin, a stain commonly used on fresh material to show up cellulose, is of little use on archaeological specimens.

Fortunately the Egyptian flora is relatively small and the same species were regularly used for the manufacture of articles, but even so a standard reference is required for comparing material under examination. The Jodrell Laboratory at Kew has a vast collection of microscope slides, accumulated over many years, first by L A Boodle and then by Dr C R Metcalfe and the present head of anatomy, Dr David Cutler, who kindly helped me with this study. These slides are permanent, stained preparations of sections of plants – leaves, stems or wood. They have usually been taken from living material, so when using them for comparison with ancient objects allowances must be made for shrinkage and distortion. Restoration of the original appearance is sometimes possible after treatment with Parazone or chlor-zinc-iodine reagent.

Some archaeological traps

Not all the plants found in excavations are ancient, or at least contemporaneous with the civilisation being excavated. For example, timber is often reused for building purposes. Any attempt to date a building by carbon dating its timber will be of doubtful accuracy unless it can be checked by independent means. Another problem is when ancient graves have been dug through, thereby introducing more recent material among older strata. The desert wind can also play tricks, by sucking in such things as cigarette packets, while rodents carry items into their holes. Sometimes obviously incongruous material is mixed with old samples. Thus I was surprised that the excavations in the sacred animal necropolis at Saqqara yielded corn (maize) cobs, which must be modern material. It so happens that Boodle also realised that he was examining maize fibres in the so-called 'dog's bed' sent to him from the cemeteries of Armant! But this calls into question less obviously intrusive material which could easily be ancient or recent. For example, in the absence of carbon dating it is difficult to be sure whether the pine cones, chestnuts and hazel nuts (which all must have come from Europe or Western Asia) also retrieved from Saqqara are contemporary with the ancient cereals and other objects from the tombs.

A few years ago there was the widely reported find of tobacco by the French Egyptologists investigating the mummy of Rameses II. Since tobacco (*Nicotiana* species) is entirely American, these finely ground particles must have been a later introduction, probably in the form of snuff during the 19th century.

Living seeds?

Probably the question most frequently asked about Ancient Egyptian plants is whether the cereals found in the tombs have ever been germinated. The short answer

is 'no!'. Popular opinion is still inclined to believe stories of mummy wheat germinating but the scientific evidence is against it. In fact, most of the seeds of the numerous species found in tombs are carbonised. There can be little doubt that any grains which have produced healthy plants were cunningly introduced into the tombs or somehow switched with ancient ones. Although the dry atmosphere of Egyptian tombs has preserved in a remarkably complete state most of the objects hidden in them, seeds rapidly become too dry to germinate. Fresh seeds are alive: their embryos are dormant and the life processes continue, at a greatly reduced rate. In order to stay alive, they need moisture, albeit a very small amount. However, even if the seeds in the tomb had not dried out, the natural processes of ageing (which are still not fully understood) would have taken effect, for a time comes when the protoplasm (the living contents of the cells) can no longer reproduce itself, and life ceases altogether.

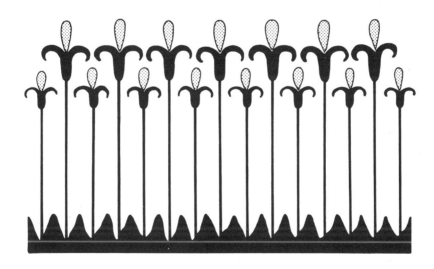

1   Flowers and Leaves

Flowers and floral decorations were an integral part of the Ancient Egyptian civilisation, both for the living and the dead. Floral offerings were found in many noble and royal tombs, and scenes of flower gardens were painted on their walls. Flowers and plants were included not only for their intrinsic beauty but also for symbolic reasons, or an account of their uses. In death they formed gorgeous garlands and bouquets.

In this chapter we look at the actual material left in Tutankhamun's tomb, and at some of the objects found which were ornamented with floral motifs or made in the shape of flowers. Here, and in the other chapters, the plants are referred to by their English name and, on their first mention, by their Latin name with a page reference to the fuller description given in the second part of each chapter.

The Antechamber of Tutankhamun's tomb as discovered by Howard Carter in 1922. The propped-up stick bouquet **on the right is of persea leaves *Mimusops laurifolia*. Note the beautifully painted casket (shown in detail in pl 27).**

Finds from the Tomb

Stick bouquets

As Howard Carter exchanged his candle for an electric torch to shine through the hole in the doorway to the Antechamber, the improved light enabled him to recognise not only the treasures but humble bouquets of leaves. Later examination showed that these were leaves of the persea tree (*Mimusops laurifolia* p 15) and olive (*Olea europaea* p 16) tied to a stick of common reed (*Phragmites australis* p 35).

One can only conjecture the reason for placing these funerary bouquets in the tomb. There were no flowers among the leaves and the bouquets were found propped up against the wall near the two life-size statues of the king guarding the entrance into the next room. As the reed would make a handle they were probably carried by members of the funeral cortège.

Floral garlands and collars

Further floral tributes were found on the coffins and were examined by Percy Newberry on the spot. They were breathtakingly beautiful even after 3,100 years, and most of the flowers were well preserved and readily identifiable. However, all of them were in a very delicate condition and some fell into dust on being touched.

A small wreath and a garland were found laid on the second coffin. On the forehead of Tutankhamun's golden mummiform coffin a wreath was looped over the gold-moulded cobra (known to Egyptologists as the uraeus) and vulture (symbols of Lower and Upper Egypt). Newberry later wrote that 'in the manufacture of this wreath a strip of papyrus pith served as a foundation; over it were folded leaves of olive which served as clasps for the cornflowers (*Centaurea depressa* p 14) and waterlily petals; the olive leaves were securely fastened together in a row by two thinner strips of papyrus pith, one placed over, the other under, alternate leaves. The leaves were arranged so that one leaf had its upper surface outwards, the next with its under surface outwards, this arrangement giving the effect of a dull green leaf beside a silvery one.'

Papyrus pith (*Cyperus papyrus* p 33) in strips was also used for the pectoral garland, which was loosely arranged in semicircles on the coffin breast. There were four bands: two with olive leaves and cornflowers, one with willow leaves (*Salix subserrata* p 17) folded as clasps to hold cornflowers and waterlily petals, and the fourth again with willow leaves, but this time with wild celery leaves (*Apium graveolens* p 14) as well as cornflowers.

A mini-wreath of olive leaves and cornflowers around the cobra (uraeus) on the golden head of the first (outer) coffin.

The greatest glory, however, was the floral collar found upon the innermost coffin around the golden face of Tutankhamun. The gorgeous colours of the fresh blue and yellow flowers interspersed with tiny blue-green faience rings, and the green leaves and red and yellow fruits must have glowed vividly in the light of the flickering wicks held by the priests as the collar was laid on the solid gold coffin. In fact, the picture of Tutankhamun and his wife on the back of his golden throne shows the king wearing such a collar (see pl 4).

The collar had nine rows of ornaments arranged on a semicircular sheet of papyrus pith. Thin strips of date palm leaves (*Phoenix dactylifera* p 62) forming the first, second, third and seventh rows carried blue-green faience rings and berries of withania nightshade (*Withania somnifera* p 18). Strips of papyrus bound the fifth row, which was of willow leaves, alternating with pomegranate leaves (*Punica granatum* p 62) and blue waterlily petals (*Nymphaea caerulea* p 16). A date palm strip was again used for the fifth row, but only to string berries of withania, while the next row incorporated cornflowers, ox-tongue (*Picris radicata* p 16), and more pomegranate leaves, with eleven half persea fruits sewn on to the collarette. Other leaves

Photo: H Burton

Photo: H Burton

Garlands of olive leaves, blue lotus waterlily petals, cornflowers and celery leaves on the second (middle) coffin.

Around the golden effigy of Tutankhamun's inner coffin Howard Carter found a broad floral collarette, which was examined by the botanist P E Newberry.

including olive formed the eighth row, and the last was similar except for the insertion of cornflowers. Newberry concluded from the flowers and fruits available for this collar that Tutankhamun was buried at some time between the middle of March and the end of April.

Floral collars were found not only inside the tomb, but outside it, in a pit near the entrance. This pit was found by Davis in 1908 and inside the jars were three complete collars and fragments of several others. The foundation of such collars was papyrus sheets sewn together, sometimes with pieces of cloth around the edges. The whole was covered with floral decoration in concentric rows using the plants already mentioned above: olive leaves and cornflowers, with withania berries strung on to strips of date palm leaf.

Ten of the statues of deities in Tutankhamun's tomb were also garlanded. These little wreaths were constructed in a similar way using a length of papyrus with leaves, petals and flowers folded and sewn into position. The same species were used – willow, olive, persea, lotus, cornflower – together with pomegranate leaves and flowers of mayweed (*Anthemis pseudocotula* p 13). All these

flowers and leaves must have been readily available fresh, as it is impossible to fold leaves that are already dry.

A garden scene with floral bouquets

An ivory casket from the tomb now in the Cairo Museum depicts on its lid a splendid garden scene with Tutankhamun and his wife holding two gorgeous floral bouquets (see cover picture). These are composed of bundles of papyrus stems inserted with lotus flowers and gilded poppies. Behind them is a pergola with a vine growing over it festooned with ripe grapes. Below their feet kneel two of their children, picking poppy flowers and mandrake fruits (*Mandragora officinarum*, p 15). Around the edge of the lid are floral designs depicting cornflowers, mandrakes and poppies, with a fish pond on the side of the casket.

This type of garden design is found in many other tombs. A typical layout has a formal walled garden with trees surrounding a pool in which water plants flourish. Such gardens were designed for pleasure, often with a lodge or seat, lines of shade trees, fruit trees and flowers for fragrance and ornament. Sometimes the gardener is shown collecting water in the bucket at one end of a shadoof with a large weight at the other end of the arm to give balance.

Lotus and papyrus motifs

The Ancient Egyptians were great artists who used botanical designs to a large extent. Two of the commonest plant motifs are the lotus waterlilies (*Nymphaea lotus* and *N. caerulea* p 16) and the papyrus sedge (*Cyperus papyrus* p 33). It is not surprising that they were used so frequently, for both were important symbols – the lotus representing Upper Egypt and papyrus Lower Egypt – and the stalks of the two plants were shown entwined to represent the union of the country that finally occurred during the 1st Dynasty. By Tutankhamun's time they were drawn in stylised form – the lotus as the enigmatic lily of the south shown on his furniture and worked into elaborate palmette designs.

There are two species of lotus in Egypt, the white and the blue, and both featured in designs throughout Ancient Egyptian history. The blue one, being fragrant, typically appears as a bouquet or a single flower held in the hand. Here should be mentioned a constant source of confusion, dating back to Herodotus' time, over the botanical identity of the 'lotus'. The indigenous waterlilies were the blue-flowered *Nymphaea caerulea* and the white *N. lotus*, which were the ones represented in ancient Egyptian art. However, these were often confused with the eastern sacred lotus (*Nelumbo nucifera*), which was introduced to Egypt from India in the Persian period, and the confusion has continued ever since. The eastern sacred lotus is distinguished by having its pink flowers

Mandrake fruits and leaves (left), red poppy flowers (centre) and blue cornflowers (right) appear as inlaid motifs on a box in Tutankhamun's tomb. From E Wilson, *Ancient Egyptian Designs*

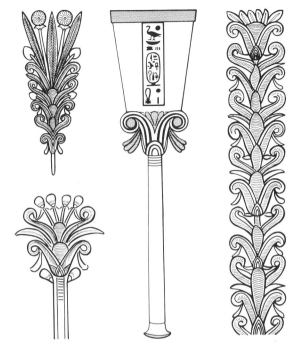

Lilies and palmettes are often depicted as ornamental designs on objects in Tutankhamun's tomb, but their botanical identity is still in dispute. The burnisher in the centre is described on p 29. From E Wilson, *Ancient Egyptian Designs*

and round leaves held high above the water surface. Its fruits are like pepperpots.

I was astonished when I saw a triple alabaster lampstand at the British Museum's Tutankhamun exhibition in 1972 (p 12), for it looked white-hot owing to the way it was illuminated: an electric light bulb was suspended in the cavity and the alabaster walls were thin enough to be translucent – originally it would have had castor oil and a linen wick in it. In 1963 I had seen it unlit

Photo: H Burton

This magnificent alabaster lamp takes the form of a three-flowered white lotus waterlily, but the side leaves are not toothed and therefore are like those of the blue waterlily.

in the Cairo Museum and even then I would have agreed with Carter, who called it 'exquisite'. It is made from a single piece of alabaster, carved with the utmost skill into the form of a white lotus waterlily (*Nymphaea lotus* p 16), with a large central flower, two smaller buds and lateral leaves shown as if they were floating on the water surface, standing 28 cm (11 in) high from a circular base.

This lamp is by no means the only object in the tomb having the lotus form. Representations of waterlilies occur time and again in Tutankhamun's jewellery (pl 7), as the capitals of columns on the alabaster ornaments (pl 3), and even on horse blinkers. One of the most remarkable lotus ornamentations is that on the two silver trumpets. These are long and slender, like Victorian coaching posthorns, with a flared bell end. The slender stem of each trumpet represents the lotus stalk, and the flared portion the flower, with the sepals and petals engraved on the outside. Moreover, each trumpet was provided with a wooden core that fitted exactly inside, with the solid bell portion decorated and coloured like a waterlily.

Papyrus is depicted in wall paintings as well as being used as a decorative motif. In pharaonic times great swamps grew in backwaters of the Nile and especially in

Photo: H Burton

Tutankhamun's silver trumpet was made in the form of a lotus waterlily, with a wooden core (right) which fitted inside the metal when not in use. The trumpet has recently been played, giving a clear, shrill sound.

the Delta. Many murals represent kings or nobles hunting wildfowl among the papyrus using throw-sticks and spears. A clump of papyrus is a favourite motif on Tutankhamun's furniture. For example, it appears on the gold-plated foot panel of a folding bedstead, on the side of

Tutankhamun's *kherp*-sceptre, a symbol of authority, was made of wood in the form of a papyrus flower head, covered in gold, and ornamented with animal sacrifices on the blade.

Photo: H Burton

a casket, and the back of a throne, as well as on the gilded shrine, where Queen Ankhesenamun is shown kneeling at the king's feet and, handing him an arrow to shoot at the ducks.

In a beautiful alabaster centre-piece in the form of a ship, the canopy pillars are in the shape of the 'papyrus columns' often seen in temples. The bulging form relates to the time before the Egyptians built in stone, when they used bundles of papyrus stems bound top and bottom to support a light awning. In the course of time this form was adopted by temple architects and stone masons, who retained the bulging shape, the fluting and the horizontal binding. Even the papyrus basal scale leaves can be seen represented on this alabaster model. Sometimes the papyrus theme might extend to the capital itself, which would then appear as an inflorescence, as can be seen in miniature in Tutankhamun's golden *kherp*-sceptre.

PLANT SPECIES

Anthemis pseudocotula
Mayweed, camomile
Family: Compositae (see pl 5)

Of the dozen or more mayweeds recorded from Egypt, *Anthemis pseudocotula*, *A. melampodina* and *A. microsperma* are the only common ones. The first occurs in the

Flowers of mayweed *Anthemis pseudocotula* and lotus waterlily *Nymphaea caerulea*, decorating one of Tutankhamun's sandals.

Photo: H Burton

Nile Valley in fields and cultivated ground, while the others grow in sandy places nearer the coast. It is likely, therefore, that the ancient Egyptian artist who ornamented Tutankhamun's chariots with mayweed flowers was familiar with the weedy species and that these were the ones placed in the floral collars (p 10).

A. pseudocotula is an annual herb with finely divided pinnate leaves on slender, erect, branched stems growing about 30 cm (1 ft) high. The large daisy flowers have white ray florets and bright yellow disc florets.

Apium graveolens
Wild celery
Family: Umbelliferae

Celery is a biennial herb 30–80 cm (1–2½ ft) high of moist places. Its large, broadly pinnately lobed leaves are borne on thick stalks, which are edible in cultivated plants, but are too strongly flavoured for eating in wild ones. The flower heads are typical of the parsley family, having an umbrella shape and numerous small white flowers.

It is likely that cultivated plants were used for the garland over Tutankhamun's mummy, the leaves probably having been chosen for the sake of their strong characteristic fragrance.

Wild celery *Apium graveolens*.

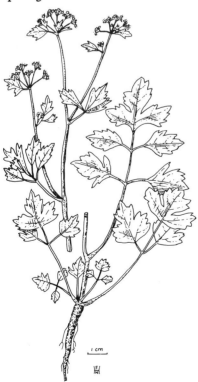

Centaurea depressa
Cornflower
Family: Compositae

Judging by the frequency with which this cornflower appears in tombs, in wall paintings and as faience models for necklaces, it must have been commonly grown as a cultivated plant. It has not been reported from Egypt for many decades and this species appears to be limited to cultivated ground in the Middle East from Turkey to Baluchistan, so it is doubtful whether it would ever have occurred naturally as a weed in Egypt, which is outside its likely range.

This cornflower is an annual or possibly a biennial, with a number of rather stout leafy stems growing to 30 or 40 cm (12–16 in) high. Each stem is topped by a thistle-like flower-head bearing decorative blue florets. The entire plant is covered with silvery hairs.

A wild cornflower *Centaurea depressa*.

Mandragora officinarum (including *M. autumnalis*)
Mandrake
Family: Solanaceae (see pl 6)

Mandrake is a member of the potato, tobacco and deadly nightshade family, so it is hardly surprising that like these it possesses narcotic attributes. Unlike them, however, it grows like sugar beet, having a tap root with a tuft of crinkled leaves about 30 cm (1 ft) long. The root can be enormous, as I found in 1985, when I tried to dig one up with a pick, and spade, helped by a group of students. Together we made a hole about 1 m (3 ft) deep and 1.5 m (4½ ft) across, and still we failed to reach the tips of the several branches of the tap root (pl 6).

During the autumn or winter months, pale purple flowers crowd the centre of the rosette. By spring a cluster of rounded or pear-shaped fruits about 2 cm (1 in) long appears, and as the fruits ripen, they turn uniform yellow and soft beneath the thin skin. The flesh has a characteristic sweet smell and is said to be edible – but it may also be narcotic and hallucinogenic. The fruit is very similar in shape to that of persea, but whereas persea has smaller reflexed sepals, the mandrake has a distinct calyx covering the lower part of the fruit. This feature should make it distinguishable from the other in murals and other art forms, but this is not always possible. Mandrake-like fruits found in excavations or tombs have always proved to be of persea; mandrake fruits themselves have yet to be seen. Representations of the mandrake plant and its fruits are frequent in Tutankhamun's tomb, as well as in others of the 18th Dynasty onwards. Scenes of gardens elsewhere often depict it growing as a cultivated plant, for it has never been a native Egyptian species, since it needs a Mediterranean type of climate with winter rainfall.

Mimusops laurifolia (synonym *M. schimperi*)
Persea
Family: Sapotaceae

This is a medium-sized evergreen tree up to 20 m (*c.*65 ft) high. The simple oval leaves are quite leathery and clustered towards the tips of the slender twigs. Among the leaves appear brown-covered flower buds and inconspicuous yellowish flowers. The yellow fruits are about the size of a pigeon's egg and contain two shiny hard seeds. As they are never available these days, I have not tasted any, which is a pity since Theophrastus the Greek botanist claimed that they were 'sweet and luscious', and easily digested in any quantity. Judging by their frequency in Egyptian graves they were popular fruits. The difficulty

Mandrake *Mandragora officinarum* fruits being picked by the pharaoh's daughter. This scene is depicted on the ivory casket shown on the cover.

in deciding whether the representations on murals and ornamentals are of persea or mandrake is discussed under 'mandrake'.

Since this species of persea naturally inhabits the hills of Ethiopia it must have been cultivated in Egypt, as its leaves have been found in other tombs and only fresh ones could have been folded for stitching into the garlands. It is surprising to know that a jar of honey from another tomb contains persea pollen as one of the two principal species (see p 50). These trees grow well in Cairo and it is unfortunate that those planted by Professor Schweinfurth beside the Cairo Museum have been removed in recent years.

Flowering and fruiting persea shoots *Mimusops laurifolia.*

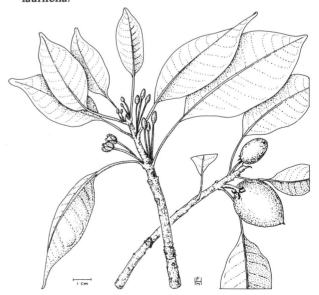

Nymphaea caerulea
Blue lotus waterlily, blue waterlily
Family: Nymphaeaceae

The margins of the floating leaves of this species are not toothed and the blue-petalled, fragrant flowers are smaller than those of the white lotus waterlily (see below). They are held just above the water on stalks 1 m (3 ft) long. A recent theory is that the fragrance is a sexual stimulus, hence the frequent depiction in murals of people smelling these flowers. Its thick creeping rhizome is sometimes used as a food in times of famine.

The blue lotus now occurs mainly in the Delta, although formerly it was common in ditches, pools and side waters of the Nile southwards into tropical Africa; it also grows in wet places in Israel but is now on the verge of extinction there.

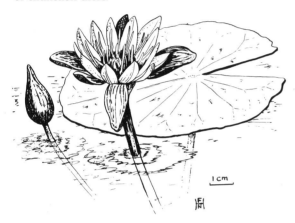

Blue lotus waterlily *Nymphaea caerulea*.

Nymphaea lotus
White lotus waterlily, Egyptian lotus waterlily
Family: Nymphaeaceae   (see pl 8)

This is the larger of the two Egyptian waterlilies, easily distinguished not only by its white petals but by the sharply toothed margins of the floating leaves, up to 20 cm (8 in) across.

The white lotus favours deeper water than the blue, but both are rare in Egypt nowadays, especially the white one. It is distributed southwards into tropical Africa.

Olea europaea
Olive
Family: Oleaceae   (see pl 9)

Since olive leaves feature in the garlands of Tutankhamun's tomb they must have been cultivated locally. The small, rounded trees are a feature of the parts of the Mediterranean region where the winters are cool and moist and the summers hot and dry. They will grow under irrigation in Egypt, even in oases, but they do not yield as well as in the rainfall areas and it is known that large quantities of olive oil were imported into ancient Egypt from Retenu, present-day Israel. Olive oil may well have featured in the jars in the tomb. Olive oil has long been used for cooking, anointing skin and hair, and as a soothing ointment with spices dissolved in it for healing wounds such as were incurred by the man on the road to Jericho (Luke 10:34)

Although the trees are slow-growing and become gnarled and hollow, the timber is excellent. Their evergreen leaves are long and narrow, darker on the upper surface than beneath. Slender inflorescences of white, four-lobed flowers with two stamens appear in spring and produce stone fruits later in the year. Olives are pickled in brine or crushed to produce the fine oil.

Papaver rhoeas
Corn poppy
Family: Papaveraceae   (see pl 10)

The well-known corn poppy is an annual with scarlet flowers. The base of the four petals is usually black and can be seen clearly when the flower is open, so it is interesting to note that Egyptian artists often depicted the poppy flowers with black spots on the *outside*. Its fruiting capsule shakes in the wind, scattering its numerous small seeds. The divided leaves and slender stems have fairly long stiff hairs all over them.

Poppies undoubtedly originated in the eastern Mediterranean region and their seeds have been carried around the world with the grains of the cereals amongst which they grow as weeds. In the tomb, poppy flowers are depicted on the lid of a casket (front cover) and in an ornamental bouquet (title page).

Picris radicata (synonym *P. coronopifolia*)
Ox-tongue
Family: Compositae

An annual plant with a rosette of basal leaves and a branched inflorescence up to 30 cm (1 ft) high bearing daisy flower heads, the florets yellow, all similar. It grows

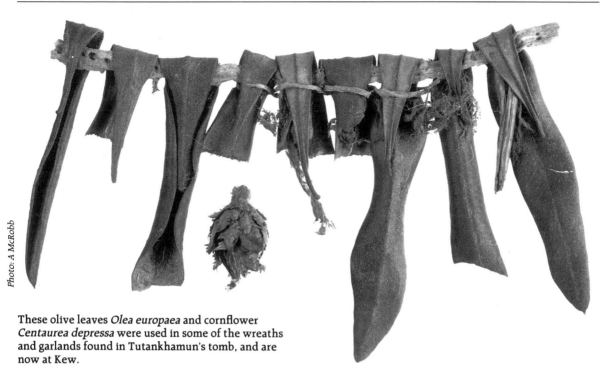

Photo: A McRobb

These olive leaves *Olea europaea* and cornflower *Centaurea depressa* were used in some of the wreaths and garlands found in Tutankhamun's tomb, and are now at Kew.

commonly in the desert sandy soil of the Nile Valley, flowering in March and April. Occasionally it was used in funerary garlands.

Ox-tongue *Picris radicata*.

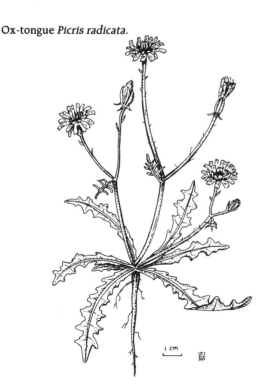

1 cm

Salix subserrata (synonym *S. safsaf*)
Willow
Family: Salicaceae   (see pl 11)

Willows are quite common along the banks of the Nile and in the Delta area, in places where their roots are always moist. This species of willow is a small tree or branched shrub. Its twigs root easily, and even the fallen trunk can sprout roots and develop new shoots as it lies in the mud. The narrow leaves are slightly toothed along the margins. Catkins on separate male and female trees develop very early in the year, when the leaves appear on the brittle twigs.

The leaves were used in Tutankhamun's garlands, but twigs were not employed in basketry as is the case with other species in Europe. The wood was not recorded from Tutankhamun's tomb, although it is known to have been used in the handle of a very ancient flint knife and a box made from its timber has been found, dating from the 3rd Dynasty.

Photo: A McRobb

Leaves of the willow *Salix subserrata* folded over strips of papyrus pith as a wreath.

Withania somnifera
Withania nightshade
Family: Solanaceae

Withania nightshade *Withania somnifera.*

The fruits in the garlands were originally identified with the woody nightshade (*Solanum dulcamara*), but this species is not found in Egypt, although it is widespread in Europe. They now are said to belong to the Egyptian withania nightshade, which also has yellow-orange fruits enclosed in an inflated calyx. I have not been able to confirm this identification, which I find surprising, since a very common nightshade, *Solanum villosum*, has similar but more conspicuous fruits.

Withania is a perennial or undershrub about 40 cm (16 in) high, with hairy stems and rather large ovate leaves. The heads of flowers occur in the leaf axils, each small flower having five reflexed greenish petal lobes. The berries develop into round yellow-orange fruits, about the size of garden peas, enclosed in the inflated calyx.

It occurs in waste places in North and Tropical Africa, and across Asia. Professor Newberry observed that only berries of this plant (which he called woody nightshade) have been found in Egyptian tombs and that they are always threaded on to thin strips of the leaves of date palm. It was obviously a very persistent practice, as it was still being done over a thousand years later in the Graeco-Roman period, and it was mentioned by Pliny as being used by the Egyptian garland-makers. The fruits of withania are well known in tropical Africa, where their alkaloidal properties are used in a variety of medicines.

2　Oils, Resins and Perfumes

Thirty-five beautiful vases were found stacked in the Annexe storeroom of Tutankhamun's tomb. Decorated with floral designs (several of them with columns carefully cut to represent stems of papyrus and lotus), most of them are made of what is popularly called alabaster, but strictly speaking is calcite. The skill of the ancient craftsmen who made these wonderful vases was probably matched by the art of the perfumers who blended their contents (see pls 12 and 13).

Altogether, these vessels would have held some 350 litres of precious oils, which must have been worth a pharaoh's ransom. One actually had a capacity of 14 litres.

A beautifully carved alabaster (calcite) perfume vase depicting the king and queen, found in the burial chamber. The sides take the form of lotus and papyrus stalks and flowers. Note the jar (shown in detail in pl 12), and the mayweed flowers ornamenting the linen pall.

Photo: H Burton

No wonder, then, that the tomb robbers who entered shortly after the burial tipped out the contents of all the vessels. One of the thieves had actually left his fingerprint inside a jar as he scraped out the remaining ointment! However, at the bottom of the largest jar some oil remained, and Carter found that this was still viscid beneath the crust. A thorough analysis at the British Museum and Kew concluded that it consisted of 90 per cent animal fat and about 10 per cent of some resin or balm.

The perfumed oils would have been used as unguents for anointing the head and body. In Egypt's hot dry climate they would have been a blessing to the skin and scalp, while creating a fragrant atmosphere for the living and the dead. About 'two bucketfuls' had been poured over the king's mummy so that when the discoverers tried to lift the body from its sarcophagus it was impossible to extract it from the mass of solidified unguents. Carter noticed that the oils had not been poured over either the head or feet of the mummy, though it reminded him of the biblical reference to Mary's anointing of Jesus' head at Simon's house with precious spikenard ointment from an alabaster jar, and Christ's reply that she was anointing his body prior to burial (Mark 14:3–9).

Finds from the Tomb

Oils

There were several readily available oils at the time Tutankhamun lived. We can only guess at which ones filled his funerary jars, but it is likely that they contained balanos oil (*Balanites aegyptiaca* p 23), safflower oil (*Carthamus tinctorius* p 32), linseed oil (*Linum usitatissimum* p 34), ben oil (*Moringa peregrina* p 25), olive oil (*Olea europaea* p 16), almond oil (*Prunus dulcis* p 62) and possibly sesame oil (*Sesamum indicum* p 27).

Oil was also used for lamps and torches. A torch-holder made of bronze and gold still had the wick of twisted linen in position where it would have drawn up castor oil (*Ricinus communis* p 26) from the cup. Several alabaster lamps that were found were in the shape of a lotus flower.

Resins and incense

Quantities of black resin fragments found in Tutankhamun's tomb were tested by Boodle and found to resemble acacia gum (p 22) in an altered condition, presumably owing to its age. True gum is soluble in water, forming a mucilaginous glue, whereas resins such as those from conifers are insoluble in water but will dissolve in alcohol or turpentine.

There were also lumps of natural resin of unknown origin, which were reddish brown and translucent. Various objects such as large beads and ear studs included in the tomb were made from resin lumps, and there was also a scarab beetle made from black resin.

A small amount of yellowish-brown incense was found which was brittle, burnt with a smoky flame and gave off a pleasant aromatic odour, according to Lucas. He deduced that it was not ladanum, mecca balsam, or storax, and probably not myrrh, bdellium or galbanum either, but was possibly frankincense (*Boswellia* p 23) made into balls. Frankincense and myrrh (*Commiphora myrrha* p 24) produce oleo-resins which are partly soluble in both water and alcohol. There was a famous Egyptian incense known as *kyphi* said to contain ten or even sixteen component substances, but their exact identity remains a mystery.

Coniferous resin can be obtained from most conifers – excluding cypress, juniper and yew – those in the area of Egyptian trade being only Cilician fir (*Abies cilicica* p 44), pine (*Pinus halepensis* p 26) and oriental spruce (*Picea orientalis*). One of these, either Cilician fir or 'cedar', yielded the famous resin known as *ash* which was named in hieroglyphics on one of Tutankhamun's jars. Several of the statues in the tomb were coated with a black varnish which has not been positively identified – some researchers consider it to be a coniferous pitch resin and others consider it to be bitumen. Coniferous resins were important for mummification. Other non-coniferous resins used for this purpose were mastic and Chios balm (*Pistacia lentiscus* and *P. atlantica*, p 26).

Perfumes

As with the oils, we can only surmise which perfumes were stowed away in those wonderful jars. In other tombs wall paintings show the preparation of perfumes from the flowers of white (or Madonna) lilies (*Lilium*

A necklace of resin beads from the tomb.

Photo: H Burton

Plate 4 On the back of the golden throne Tutankhamun and his Queen Ankhesenamun are shown wearing floral collars similar to those found in his tomb. Beside them are gorgeous floral bouquets with papyrus, lotus and poppy flowers. See also pl 28.

Plate 5 (*top left*) Mayweed *Anthemis pseudocotula*.

Plate 6 (*above*) Mandrake *Mandragora officinarum* roots are stout and deep, as the author proved while trying to dig up one plant.

Plate 7 Tutankhamun's pectoral with semiprecious stones set in gold in the form of lotus, poppy and papyrus.

Photo: F N Hepper

Plate 8   White lotus waterlily *Nymphaea lotus* at Kew.

Photo: F N Hepper

Plate 10   Corn poppy *Papaver rhoeas*.

Plate 9   Olive *Olea europaea* growing in a small garden beside St Catherine's monastery in Egyptian Sinai.

Photo: Robert Harding Picture Library

Plate 11 (*top left*) Male catkins of the willow *Salix subserrata* at Aswan. Willow leaves were used in Tutankhamun's wreaths.

Plate 12 (*above*) This calcite cosmetic jar still contained some fragrant substances when found in Tutankhamun's burial chamber.

Plate 13 Tutankhamun pouring fragrant oil into the hand of Queen Ankhesenamun. From the gilded side of the Nekhbet shrine.

candidum p 25), which must have been grown in special gardens, since they originate in the moister eastern Mediterranean region. Other flowers such as henna (*Lawsonia inermis* p 25), as well as slivers of coniferous timber, fragrant bark, resins, and sweet-smelling desert grasses and herbs, were used. They were steeped in oil which took up the fragrance, and wall paintings show how a squeezing tourniquet was applied to the bag in which the material was placed over the jars. Some detailed recipes are provided by later authors, such as Theophrastus, Dioscorides and Pliny, who are quoted by Manniche in *An Ancient Egyptian Herbal* (see Further Reading). In Old Testament times the sacred anointing oil was composed of a rich blend of fragrant spices (Exodus 30: 22–25).

Adhesives

Resin was mixed with whiting to fasten inlays of stone and glass on the coffins and caskets. In Tutankhamun's tomb there were small tears (2–5 × 0.5 mm) and rods mixed with natron which have not been identified, although both frankincense and myrrh have been ruled out. It is known that acacia gum was probably mainly used for gluing together small pieces of papyrus, linen and suchlike, but animal glues were required for joinery. In this tomb a black gum-resin was mixed with lime and sand to repair the broken lid of the stone sarcophagus.

A sample of red very powdery resin found there was tested by H J Plenderleith of the British Museum Research Laboratory. It was 93.9 per cent soluble in alcohol, 6.1 per cent in water and it smelled of 'terebinth' on being heated. His suggestion that it was 'Angola copal' is rather doubtful, as this is a central tropical African resin, and mastic (p 26) is more likely.

Mummification materials

By an extraordinary chance the embalming materials for Tutankhamun's body were found in 1908 by Theodore M Davis. The materials were found in jars in a pit dug out of the stone floor of the valley and covered up with debris. The mummification process inevitably produced a good deal of unpleasant refuse, which was considered too unclean to be put into the tomb, but because it had been in contact with the royal body it could not be thrown away. Such material was therefore packed into large jars and buried separately from the tomb, though often not far away. Many of these 'embalmers' caches' are known, but Tutankhamun's is the only royal one to have been found. However, when Davis discovered it he did not realise its

significance and it was some thirty years later, in 1941, that H E Winlock wrote up a complete account of the finds. Objects of botanical interest included numerous linen bandages and other cloths, as well as several dozen bags filled with chaff and natron – the soda and salt essential for embalming. There were also some trays of mud thought to be 'Osiris beds' (see p 54). Davis was excited to find a little head of plaster and linen, like a miniature mummy mask, of unknown significance. Reed sticks, two pieces of papyrus stalk and some wooden sticks were also present. Jar stoppers made of halfa grass (p 33) were found and there were fragments of two circular papyrus jar stands. Inscriptions on small offering cups showed what they had been intended for: 'incense[?] for fumigation', 'a drink', 'half-*pk* loaves' and 'grapes'. There were also wine jars and bottles, as well as numerous other vessels and dishes, and remnants of a banquet of geese, ducks, and joints of sheep and cow. Two brooms, one of reed, the other of grass, would have been used to tidy up after the burial. Of particular interest to us are the floral collars Davis found in several jars (see p 10).

Mummification is a very ancient practice which was continued in Egypt for several millennia. Various resins, oils and spices were used for mummification, which embalmed the body to save it from decay, but we cannot go into the complicated rites and processes here.

Embalming is a word derived from the Latin for balsam, which may have been found in this tomb, while the words mummification and mummy seem to have come from a Persian one for bitumen – although it is doubtful whether this substance, obtained from the ground, was used, while the resins from plants, which look similar, certainly were.

Tutankhamun's body was embalmed in the traditional way with the soft internal organs removed for preservation in four jars which were housed in a special gilded box (the canopic shrine). After being washed in water and oiled, Tutankhamun's corpse was probably preserved with the natron and salt mixture, the residue of which was found outside the tomb. However, D E Derry, who reported on the mummy for Carter's *Tut-ankh-Amen*, found only a few crystals of salt. Finally, the mummy was wrapped in sixteen layers of linen bandages among which were included juniper berries (p 60). It was then placed inside a series of three gold coffins, a stone sarcophagus and four gilded boxes (called shrines).

It is interesting to note that the seventy-day period of mourning accords well with the biblical account of Jacob's death: 'and Joseph commanded his servants the physicians to embalm his father. So the physicians embalmed Israel; forty days were required for it, for so

Photo: H Burton

The second shrine in the form of a southern temple, with incised reliefs in the golden covering referring to burial ceremonies.

many are required for embalming. And the Egyptians wept for him for seventy days' (Genesis 50: 2–3). Eventually Joseph, too, died 'and they embalmed him, and he was put in a coffin in Egypt' (Genesis 50: 26).

PLANT SPECIES

Acacia species
Acacia
Family: Leguminosae   (see pls 14 and 35)

Gum is obtained from the stems of several species of *Acacia* inhabiting sandy wadis and growing beside the Nile. Timber from the native acacia trees was always in demand, for boat-building in particular. Although a dozen or more acacias are listed from Egypt, most of them are rare or are shrubby, leaving about four tree species, which are described below – *Acacia tortilis*, *A. raddiana*, *A. nilotica* and *A. albida*.

A. tortilis and *A. raddiana* (often combined as sub-species of *tortilis*) are flat-topped trees, sometimes bushy, with a number of red-brown trunks, often about 5 m (16 ft) high or much taller. The compound leaves have numerous tiny leaflets, with a pair of short stipular thorns at the base of the stalk. Globular heads of whitish flowers give rise to woody pods that are curved or twisted, containing half a dozen seeds.

A. nilotica, the Nile acacia, inhabits moist places beside the river, as its name implies. It is a rounded tree about 8 m (27 ft) high with dark bark and long sharp stipular thorns. The globular flower heads are yellow and the pods are like strings of flattened beads.

A. albida, the white acacia, is unusual because it is deciduous in the summer instead of during the winter like the other species. It is a medium to tall tree with characteristic white bark, now occurring only in the Aswan area of the Nile Valley on the drier parts of the river bank. Its fragrant white flowers, borne in an elongated inflorescence, produce thick pods of an orange colour.

Gum was probably obtained mainly from the Nile acacia – true gum arabic comes from *A. senegal*, a tropical African species. Wounds on the trunk and branches yield tears of yellowish gum which turns a darker colour, even black, with age, such as that found in Tutankhamun's tomb. The gum was a well-known component of medicines, as were the boiled leaves and flowers. Acacia bark provided an important source of tannin for the preparation of leather from hides, and a blue dye for linen cloth was extracted from acacia pods (p 30).

Acacia timber is reddish, hard and durable. Locally it was used for boat-building and general joinery, so it is likely that some of the red wood reported from Tutankhamun's tomb is acacia. One dowel from the shrines has been identified as acacia wood.

Balanites aegyptiaca
Egyptian plum
Family: Balanitaceae (see pl 15)

The Egyptian plum or false balsam yields the medicinal balanos oil. This very spiny tree is abundant in the drier parts of tropical Africa and in the Tihama of western Arabia, as well as in the Arabah valley near the Dead Sea. It is usually about 3 m (10 ft) high with quite a stout trunk and a tangle of weeping branchlets, and has sharp thorns in the axils of the leaves, each of which bears two grey-green leathery leaflets. The small greenish flowers also occur in the leaf axils and the young fruits are like large acorns – hence the botanical name, from the Greek *balanos*. The plum-like fruit develops a very hard kernel which yields the oil on crushing. The yellow to brown timber, which is hard and heavy, is used for small wooden objects on farms in Africa and it was also recorded from the Cheops boats.

Leaves, flowers and fruit of the Egyptian plum tree *Balanites aegyptiaca*.

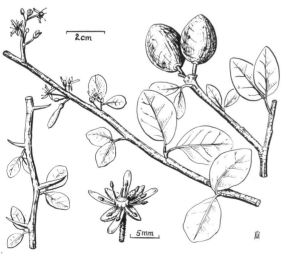

Boswellia species
Frankincense trees
Family: Burseraceae (see pl 16)

Frankincense resin was so important an incense in antiquity that it would be unthinkable for it not to be included in an Egyptian royal tomb. One is reassured, therefore, to note Lucas's analysis of some yellow-brown balls as having characteristics of frankincense. When fresh the resin is whitish. It was obtained from several species of *Boswellia* tree. *B. sacra* (including *B. carteri*) grows in the Dhofar region of Oman in southern Arabia, and in Somalia in the Horn of Africa, on the dry but misty mountains there. Both locations would have been reached by expeditions down the Red Sea such as that commemorated in Queen Hatshepsut's (*c.*1470 BC) mortuary temple at Thebes. Another species of frankincense tree, *B. papyrifera*, has a wider distribution, in the dry mountains of Ethiopia, and it is possible that its resin could have been

Flowering, leafy shoot of the frankincense tree *Boswellia sacra*.

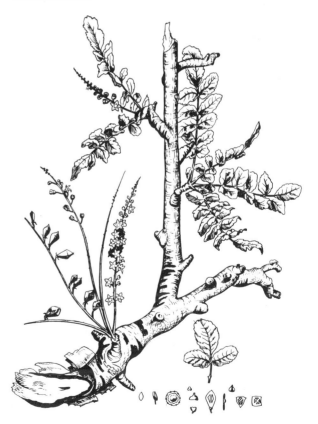

obtained by land or down the River Nile. This is the resin still used in Coptic churches.

Cuts on the trunk, where the thin bark peels off like paper, produce tears of resin. Both species have compound, pinnate leaves, but the flowers of *B. sacra* are white, while those of *B. papyrifera* are pink and in a much laxer inflorescence. The fruits of *Boswellia* are dry capsules (unlike the berries of *Commiphora* species mentioned below).

Frankincense resin was burnt in religious rites in Egypt, as elsewhere, including the Israelite tabernacle (Leviticus 16: 12), and it was brought to the infant Jesus (Matthew 2: 11).

Commiphora gileadensis (synonym *Balsamodendron opobalsamum*)
Balm of Gilead
Family: Burseraceae

The word 'balm' is loosely used for various ointments, including the balm of Gilead (which is also called Mecca balsam and opobalsam) obtained from *Commiphora gileadensis*. This is a tropical shrub traditionally said to have been grown by Solomon at En Gedi on the Dead Sea, which has a tropical climate similar to its native area in south-west Arabia, and to have been brought to Jerusalem by the Queen of Sheba (1 Kings 10: 10). It is a non-spiny shrub or small tree, with numerous slender branches bearing small compound leaves with three leaflets. These are present for a short time after the rainy season. The gummy oleo-resin is obtained by making incisions in the bark, and it is collected in the same way as the exudates of myrrh and frankincense. Ancient writers considered it a cure for many diseases and Pliny noted its extraordinary sweetness. It was often used in liquid form – made by dissolving the lumps in warm olive oil and straining the solution.

Commiphora myrrha
Myrrh
Family: Burseraceae

The myrrh bush grows about 2 m (6 ft) high on steep rocky hills in the semi-desert countries of Somalia and the Yemen. It is thorny, with long stout branch thorns which protrude in all directions, bearing small three-lobed leaves, small whitish flowers and beaked berries, although for most of the year the bush stands gaunt and leafless. A thin papery bark peels naturally from the stems, exposing a thicker green bark, which contains the fragrant myrrh.

Balm of Gilead *Commiphora gileadensis*.

Exudation of it is encouraged by cutting the stem, and as in ancient times, the reddish tears of aromatic resin are still collected by local people into baskets for sorting, before they are sold to incense dealers and transported to the markets of the world.

In Ancient Egypt, myrrh was widely used as a medicine and for perfumes, incense and for funeral rites; it was evidently less usually burnt than frankincense and more likely to be used in pink powder form.

Egyptians gathering cultivated white lilies *Lilium candidum* for perfume. From a relief of the 26th Dynasty in the Louvre, Paris.

Lawsonia inermis
Henna
Family: Lythraceae (see pl 17)

Henna is included on account of its use in Ancient Egypt for staining the nails and hair of mummies. However, there is no direct evidence that Tutankhamun's body was so treated, and there is still dispute among Egyptologists concerning its prevalence.

The shrub from which the dye is obtained is usually spindly, about the height of a man, and with several stems. A profusion of slender branches carry the leaves which are crushed to make the paste used for henna dye. The flowers of henna are white and beautifully fragrant.

In Egypt and other warm countries it is grown as a cultivated plant, but I have seen it growing wild on rock outcrops in Kenya.

Myrrh *Commiphora myrrha.*

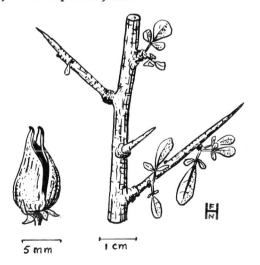

5 mm   1 cm

Lilium candidum
White lily
Family: Liliaceae (see pl 18)

If the perfume of the white or Madonna lily was included in Tutankhamun's vases, it would have come from plants specially grown for the purpose, since this is a species inhabiting hills in the eastern Mediterranean area. The cultivation of lilies for perfume is shown on some of the ancient wall reliefs and vast quantities of flowers must have been required for the perfume recipes.

The white lily has a bulb comprised of a number of scales similar to the 'cloves' of a garlic bulb. It puts forth its leaves as a basal tuft in winter long before the flower stem arises. This can be up to 1 m (3 ft) tall, bearing leaves of decreasing size towards the terminal cluster of trumpet flowers. These are white and beautifully scented.

Moringa peregrina (synonym *M. aptera*)
Horseradish tree
Family: Moringaceae (see pl 19)

In Ancient Egypt ben oil extracted from the horseradish tree was greatly used for cosmetics and cooking. This tree belongs to its own family, which is closely related to the peas and beans. It has a dense crown of numerous whip-like branchlets bearing slender leaf stalks and scattered small leaflets. The pinkish flowers give rise to long narrow pods containing three-cornered seeds about the size of a hazel nut, which are crushed to extract their oil. Ben oil is sweet and odourless, hence its use in perfumery.

This tree is native in the Jordan Valley and southwards to the Sudan. 'Ben oil' is often mentioned in works on Egypt as having been obtained from another species, *M. oleifera* (*M. pterygosperma*), but that is a native of Sudan and may not have been available then.

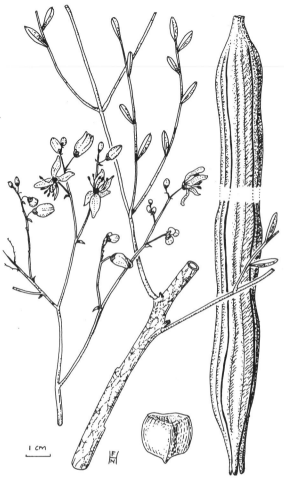

A leafy shoot, flowers, fruit and a seed of the ben oil tree *Moringa peregrina*.

Pinus species
Pine
Family: Pinaceae

Pine trees, in common with many conifers, have resin ducts in their timber. Some species are commercially tapped for resin by making vertical grooves and collecting the resin in cups attached to the trunk – turpentine may also be collected. The fossilised resin is known as amber, and it was sought after for carving into jewellery. Pine timber was also imported into Ancient Egypt, even in Tutankhamun's time – the doorway into the burial chamber had beams of pine, as well as date palm and exotic ebony.

Several species of pine tree grow in the eastern Mediterranean area, where the climate is moister than in Egypt. Those that concern us here are the Aleppo pine (*P. halepensis*), the Brutian pine (*P. brutia*) and the stone pine (*P. pinea*) – the latter yielding edible seeds that have been reported from Ancient Egypt.

These pines are tall trees, having a single trunk covered with furrowed, resinous bark and soft white timber. The leaves are long needles carried in pairs – other species have three or five together. Pine cones are the fruit, bearing the naked seeds between the woody scales.

Pistacia species
Mastic and Chios balm
Family: Anacardiaceae

The appearance of the species yielding these resins is very different: *P. lentiscus* is a low evergreen shrub, while *P. atlantica* is a large deciduous tree. Both resins were used in ancient times and still are to a certain extent, but they have not been definitely recorded from Tutankhamun's tomb, unless they were in the 'red powder', or used in the mummification process (p 21).

The mastic bush (*P. lentiscus*) is common in the maquis type of Mediterranean vegetation, where it tends to sprawl in large clumps. The leaf-stalks are winged and bear several leathery leaflets on the pinnate leaf; the small flowers produce little round fruits which turn red. Remarkably, the resin-yielding form of this species occurs only on the Island of Chios, off the Turkish coast.

Also in Chios, as well as on Cyprus and widely round the dry parts of the Mediterranean, grows the large terebinth (*P. atlantica*), which yields Chios or Cyprus balm or turpentine – confusingly, authors formerly attributed this balm to the plants of *P. lentiscus* growing on Chios.

Ricinus communis
Castor oil plant
Family: Euphorbiaceae   (see pl 20)

There is no direct evidence that castor oil occurred in Tutankhamun's tomb, but it was probably used in the torch and lamps (pp 19–20). Its use for lighting was mentioned by Herodotus and Strabo, and its medical uses were mentioned in an Egyptian papyrus. Dried seeds were found in graves dating from the pre-dynastic period and it is still commonly found on the banks of the Nile and on rubbish heaps.

The plant grows rapidly into a soft-stemmed bush and tall shrub with large palmate leaves. Its flowers develop on erect inflorescences and have no petals. They set fruits which are spherical and covered with prickles.

Photo: F N Hepper

A Chios balm tree *Pistacia atlantica*.

Castor seeds are like fat beans, mottled on the surface and oily within – they are poisonous if swallowed. The oil, extracted by crushing, has a strong smell.

Sesamum indicum
Sesame
Family: Pedaliaceae

There is considerable difference of opinion among Egyptologists as to whether the sesame and its oil mentioned in the texts was the plant we now know by that name. Some have suggested it was either hemp (*Cannabis sativa*) or linseed oil from flax (p 34). Very few records of *Sesamum* seeds have been recorded in Egypt, so it is of particular interest that a sample from Tutankhamun's tomb was noted by Boodle as 'possibly a species of this genus' and 'resembling seeds of sesamum in shape'. They are much broken, but if confirmed, this will be the earliest find in Ancient Egypt.

Botanists are also uncertain about the origin of *Sesamum indicum*, since in spite of its name (referring to India) it may have been a tropical African species. It is a herbaceous plant usually about 1 m (3 ft) high, more or

4 cm

A leafy shoot with flowers and seeds of the castor oil plant *Ricinus communis*.

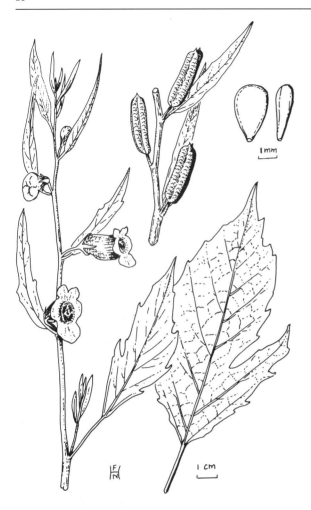

less branched. Its lower leaves are sometimes divided, with the upper ones long and narrow. The large pink or white tubular flowers occur in a long inflorescence, producing narrow erect beaked pods about 3 cm (1¼ in) long, containing the small oily seeds. The oil was used for cooking, in lamps and for medicines; the whole or ground seeds were eaten on bread.

          *   *   *

For other oil-producing plants, see: safflower (*Carthamus tinctorius* p 32), flax (*Linum usitatissimum* p 34), Levant storax (*Liquidamber orientalis* p 47), olive (*Olea europaea* p 16) and almond (*Prunus dulcis* p 62).

Leaves, flowers and pods from the sesame oil plant *Sesamum indicum*.

3 Papyrus, Flax and other Fibrous Plants

Several fibrous plants had a number of important uses in Ancient Egypt. Probably the most celebrated one was the papyrus sedge (*Cyperus papyrus* p 33) and it is not surprising that it occurred in Tutankhamun's tomb in several forms: as short lengths of unknown significance, as a box, as paper and as an art form (see pp 11–13).

Finds from the Tomb

Papyrus writing material

Although papyri were so extensively used in Ancient Egypt, only three pieces have been reported from Tutankhamun's tomb, two of them plain sheets used to support floral collars. The third one, bound on the royal mummy together with amulets and symbols, was badly decayed, but there were the remains of white linear hieroglyphs written on it forming part of the ritual of the Book of the Dead. Since it included reference to Osiris and Isis the conclusion was made by Carter that it referred to the magic associated with those amulets.

Cultivation of the papyrus plant was no longer required when the demand for paper made from its pith ceased during the later Roman period. However, in recent years the plants have been reintroduced into backwaters of the Nile to supply the tourist industry with modern papyrus writing material. Traditionally this was made by slicing the stalks longitudinally, discarding the outer green rind, and laying the slices of pith side by side, with another layer on top at right angles. These were gently beaten and pressed together, the two layers adhering on drying to form sheets of paper. The surface was then burnished with a special instrument. Some beautiful examples of these, made from ivory, and with handles in the form of a papyrus column, were found in Tutankhamun's tomb. The white pith of the papyrus stalk was known to the Greeks as *byblos*, hence the word *bybla* for books that were made of papyrus, from which is derived

Photo: H Burton

Ivory and gold writing palettes and a pen-case in the form of a column with a date palm capital. In the middle is an ivory papyrus burnisher with a capital in the form of a lily-palmette motif, which was used to finish sheets of papyrus by providing a smooth surface for writing.

29

our word 'Bible'. The Greek name for the plant, *papyros*, gave us the English world 'paper'.

Pen-cases and writing outfits

Writing on papyrus sheets was by means of rush pens using carbon ink – the carbon being obtained from the soot of cooking pots, mixed with water and gum arabic. Fortunately carbon, being an element, does not fade like modern ink made with dyes. Pens made from the rush *Juncus arabicus* (p 34) were chewed by the scribe to separate the fibres into a small brush; while pens of reed (*Phragmites australis* p 35) used during the Graeco-Roman period were cut with a penknife to form a point and a split was made to allow the ink to flow.

Palettes and pen-cases (or holders) of ivory were included with Tutankhamun's effects. A tubular pen-case in gold was made in the form of a hollow column having a capital with a palm-leaf motif (see p 29).

Papyrus boats

A golden statue of Tutankhamun standing on a model green skiff of papyrus was included among his funerary items. A passage in the Bible refers to such boats: 'There is a land of sailing ships, a land beyond the rivers of Cush which sends its envoys by the Nile journeying on the waters in vessels of reed [i.e., papyrus]' (Isaiah 18: 1–2).

Boats of various sizes with curved prows were constructed of bundles of papyrus stalks – in 1970 a replica of one, the ocean-going vessel *Ra*, was made for Thor Heyerdahl and six others to cross the Atlantic Ocean. Small personal skiffs are still constructed and used in the sudd area of the upper Nile and on Lake Chad, where in 1969 I poled one along while standing up. It was a strange experience, as the craft was very buoyant and water oozed up between the stalks around my feet when the load was too heavy. This buoyancy is due to the internal cellular structure of a pith, which has a network of loose cells trapping myriads of tiny air pockets. These act as buoys which very gradually become flooded when immersed in water.

Flax, linen and dyes

Rolls of linen cloth, a flax bowstring, tapestry woven gloves, coloured embroidery, Tutankhamun's own tunics and other items were found in the tomb. Linen literally accompanied Egyptians from the cradle to the grave. All were made from the flax plant (*Linum usitatissimum* p 34), which was cultivated extensively in Ancient Egypt.

Photo: H Burton

A small golden model of Tutankhamun standing on a green skiff, which would have been made of papyrus stalks tied together.

The cultivation of flax in Egypt is mentioned several times in the Bible. For example, during the plagues of Egypt the crop was ruined by hail (Exodus 11: 31). Later, Isaiah prophesied that when the Nile dries up 'the workers in combed flax will be in despair' (Isaiah 19: 9). Evidently Egyptian linen was used for coloured bed coverings in Jerusalem (Proverbs 7: 16) and for sails of the ships of Tyre (Ezekiel 27: 7).

Between the outer and second shrine of Tutankhamun's tomb there was a delicate pall, which was later identified at Kew as being composed of linen thread. Samples of other linen fabrics found there were studied in 1937 by R Pfister. The Egyptians used various techniques such as warp-weave and embroidery, examples of both having been recovered from Tutankhamun's tomb. The fabric was dyed red with safflower (*Carthamus tinctorius* p 32), yellow with madder (*Rubia tinctorum* p 35) and blue with acacia (*Acacia* p 22), according to H E Winlock.

Photo: F N Hepper

Two papyrus skiffs on Lake Chad.

Baskets and mats

Baskets of various shapes and sizes were found in excellent condition in Tutankhamun's tomb (see p 60). Altogether there were 116, most of them containing the offerings of seeds and food that were left there for use in the afterlife. Oval baskets with or without a lid were the most frequent, but there were also several bottle-shaped ones.

All were constructed of 'coiled work', rather than plaited as modern baskets are. Coiled work requires two elements: the cylindrical fibrous core of grass or rush, and the binding, which is wrapped around the core. The bound core is then coiled round in successive layers to build up the basket, and the layers are sewn together to hold them in position. The binding or wrapping material was usually of palm leaves torn into narrow strips.

Papyrus was used for box-like baskets of a different construction and a beautifully made one was found in Tutankhamun's tomb. For this only the tough green rind of the papyrus stem was used. Papyrus rind may have been used for the 'bitumen and pitch'-covered cradle in which Moses was hidden (Exodus 2: 3). This basket was floating among the Nile rushes, which are more likely to have been reed-mace (*Typha domingensis* p 36) than papyrus itself (*Cyperus papyrus* p 33). Reed-mace and various sedges were used in the manufacture of mats and several made of papyrus, rush (*Juncus* sp. p 34) and linen-with-rush were found in Tutankhamun's tomb. There is also a biblical reference to the flat food baskets often found in other tombs, when the pharaoh's imprisoned baker dreamt of three bread baskets (Genesis 40: 16, 17).

Photo: H Burton

The pharaoh's carefully stitched linen glove.

Sandals

Several pairs of sandals and slippers, beautifully made using beads, leather, sheet gold (even embossed with imitation rush-work), or leather and gold, with delightful floral ornamentation of lotus and mayweed were found, but it is those made of fibrous plant material that particularly interest us. The soles of one pair are woven from stems of rush (*Juncus* sp.), while the toe-strap is of papyrus rind twisted round a core. In another pair the soles are of papyrus with rush binding along their edges.

Photo: F N Hepper

Above: a sandal from Tutankhamun's tomb, incorporating doum palm leaves and papyrus rind.

Below: Ancient Egyptians making rope from flax. From J G Wilkinson *Manners and Customs of the Ancient Egyptians*, London, 1878.

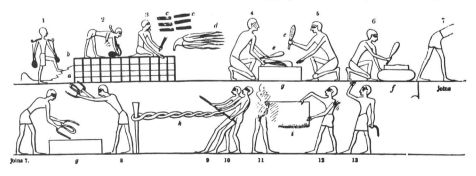

String and ropes

Mention is made on p 43 of a bowstring composed of linen thread (flax) that was found, but coarser ropes have not been reported from Tutankhamun's tomb. Egyptian ropes were composed of halfa grass (*Desmostachya bipinnata* and *Imperata cylindrica*), palm leaves (*Hyphaene thebaica* p 59 and *Phoenix dactylifera* p 62) and papyrus rind (*Cyperus papyrus* p 33). When one considers the size and weight of the stones used to build the temples and pyramids, the strength of the ropes needed to pull them into position must have been very great. The strength and thickness of the rope depended on the number of twisted component strands. Sometimes such ropes have been found discarded in some dark corner of a temple and they prove to be as thick as a man's arm. The model boats in the tomb were rigged with ropes which, in real ships, would have been made of fibrous material from the plants listed above.

PLANT SPECIES

Carthamus tinctorius
Safflower
Family: Compositae   (see pl 21)

This is a cultivated plant that tolerates sandy places. Its origin is obscure as it is not known in the wild state, although its nearest relatives occur in Ethiopia. Whole seeds were found in Tutankhamun's tomb – fresh seeds yield an edible oil which was also used medicinally.

Safflower is an annual herb growing up to about 1 m (3 ft) high with prickly leaves and a thistle-like head of yellowish-orange florets. Remarkably, these florets yield two colours – yellow and red – but the yellow colour can be washed out of cloth. The red dye, however, is permanent, unless it comes into contact with alkali. Mummy wrappings were often dyed with this plant.

Plate 14 (*top left*) White acacia *Acacia albida* beside the Nile.

Plate 15 (*above*) Balanos oil tree *Balanites aegyptiaca*.

Plate 16 A frankincense tree *Boswellia sacra* growing in a cleft of a rock in Somalia.

Plate 17 Flowering shoot of henna *Lawsonia inermis*, which is very fragrant. Henna leaves provided a yellow dye used to stain the nails and hair of mummies.

Plate 18 (*top right*) White or madonna lily *Lilium candidum*, which was cultivated in Ancient Egypt for perfume.

Plate 19 The ben oil tree *Moringa peregrina*, at En Gedi.

Plate 20 A large castor oil bush *Ricinus communis* growing by the Nile at Aswan.

Plate 21 The flowers of safflower *Carthamus tinctorius* provided a cloth dye.

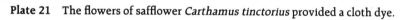

Photo: F N Hepper

Photo: F N Hepper

Plate 22 (*top left*) A flower-head of halfa grass *Desmostachya bipinnata* in Egypt.

Plate 23 (*above*) A plant with flower-heads of the imperata halfa grass *Imperata cylindrica* in Egypt.

Plate 24 A papyrus swamp *Cyperus papyrus* in Sri Lanka.

Photo: F N Hepper

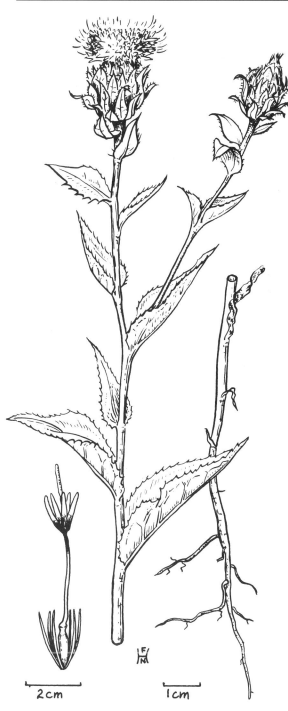

2 cm   1 cm

A plant of safflower *Carthamus tinctorius*.

Cyperus papyrus
Papyrus sedge
Family: Cyperaceae (see pl 24)

This is a marsh-loving plant that grows in enormous stands in shallow water, where its thick horizontal stems (rhizomes) root in the mud. The leaves are reduced to papery scales at the base of the triangular stalk that may attain 5 m (15 ft) or more in height, topped by a huge mop-like flower head. In winter this head is composed of thread-like green bracts, but as the warmer weather comes the small brown flowers appear.

During pharaonic times the Nile Delta supported extensive papyrus swamps, which also occurred with reeds in the Bitter Lakes region – the Sea of Reeds (the Red Sea of Exodus 13: 18). These swamps would have supplied papyrus stalks for the manufacture of the writing material. The swamps have gone, owing to pressure of population and land drainage, as well as the extraction of fresh water from the Nile for irrigated agriculture.

Desmostachya bipinnata
Halfa grass
Family: Gramineae (see pl 22)

This is a perennial grass with rigid leaves and a narrow inflorescence up to 40 cm (16 in) high. It grows in waste ground and beside canals in Egypt.

Because it is readily available and the leaves are tough, halfa grass was used for items such as mats, ropes and bags. A bundle was found in Tutankhamun's tomb and was at first thought to have been used for writing, but this was incorrect as rushes were used for pens and the halfa grass was evidently misplaced.

Imperata cylindrica
Imperata halfa grass
Family: Gramineae (see pl 23)

There are at least two grasses in Egypt known as halfa, *Desmostachya bipinnata* (above) and this one, which I am calling imperata halfa grass. They are both common and both used for the same purposes – ropes and simple ties, baskets, matting and suchlike. Only *Desmostachya* has been identified from Tutankhamun's tomb, but *Imperata* may have occurred there and is occasionally found in tombs.

Imperata is commonly found in a wide variety of habitats where water is not far from the surface. The rhizome is hard and scaly, spreading very easily and rooting from the smallest piece, hence the difficulty farmers encounter in trying to clear it from cultivated land. On the other hand, it binds sandy banks effectively. Its tough linear leaves are used for fibrous objects. At almost any time of year, especially in May, the white, cylindrical inflorescence may be seen. It stands 30 cm (1 ft) or more high.

Juncus arabicus
Rush
Family: Juncaceae

A densely tufted perennial with cylindrical stems almost 1 m (3 ft) high. Each stem contains white pith, and ends in a very sharp point. The brown flowers are found in groups to one side of certain stems, making small clusters. This rush grows in salty places in the desert. It is sometimes known as *J. rigidus* or as *J. maritimus* var. *arabicus*.

Rushes were often used for making mats and sandals and also in pharaonic times for pens, and some rush pens were found in Tutankhamun's writing cases. No doubt any other available and suitable species, such as *J. acutus*, could also be used for these purposes.

Linum usitatissimum
Flax
Family: Linaceae   (see pl 25)

The annual flax plant grows up to 1 m (3 ft) with pale blue flowers. The variety used for flax is sown close together to encourage tall-growing plants, whereas the oil seed (lin-seed) variety is sown thinly, to allow the plants to branch and flower freely. The stem fibres are extracted by the process called retting, which takes place in water after the plants have been pulled up by the roots. Retting is a bacterial action causing the breakdown of the thinner stem cells, leaving the thick phloem fibres to be combed out and dried, ready for spinning into thread. The cell fibres are each 20–40 mm (¾–1½ in) or more long.

Cloth woven from linen thread was widely used in Ancient Egypt – cotton (*Gossypium* species) was not known there in those days and not introduced until Roman times, having spread from India. Linen is cooler and cleaner for garments than wool. Flax was one of the earliest fibres known, and in Egypt there is evidence that neolithic man used it. Some of the Egyptian tombs have wall paintings showing its cultivation and the processes of retting, combing, spinning and weaving.

The Arabic rush *Juncus arabicus*, which was used by scribes.

1 cm

Common reed *Phragmites australis.*

Phragmites australis (synonym *P. communis*)
Common reed
Family: Gramineae

Abundant in marshy and salty areas of Egypt, this tall grass spreads by long rhizomes to form extensive reed swamps. Its bamboo-like canes were useful for light constructions, arrow shafts (pp 43–4) and pens. Later in the year its light brown plume-like inflorescences bend in the wind and the rough grassy leaves rustle characteristically. The species is found in wet places almost everywhere around the world.

A similar species is the giant reed, *Arundo donax*, which tolerates drier situations, such as canal banks, and develops stouter stems, with white plumes. This probably reached Egypt from southern Europe in later times, so some of the early records for it may in fact refer to the common reed.

Rubia tinctorum
Madder
Family: Rubiaceae

The madder plant is a climbing perennial herb with rough stems up to 1 m (3 ft) long. The leaves occur in whorls of four or six, and their margins are very rough, owing to the

Madder plant *Rubia tinctorum.*

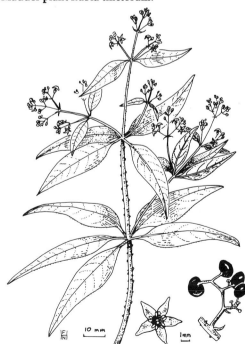

presence of numerous minute prickles. The flowers are borne in loose heads longer than the leaves and arise from their axils. Each flower is four-lobed and brownish-yellow.

Madder is typical of the maquis along the Mediterranean coasts where there is a good deal of winter rain – which excludes Egypt from its distribution. Since ancient times the roots of madder have been used as a red dye for cloth, and its occurrence in Tutankhamun's tomb was reported by Pfister as the dye in one of his fabrics, for which it must have been imported into Egypt.

Typha domingensis (synonym *T. australis*)
Reed-mace
Family: Typhaceae

This is a perennial which forms large stands in shallow water. The long erect leaves are only about 1 cm (½ in) wide. The inflorescence is very distinctive, with brown cylinders of separate male and female flowers on an erect stalk about the height of a man. The upper male inflorescence is separated by 1–2 cm (½–1 in) of stalk from the lower female part. The leaves are widely used for mats and baskets. It is also called cat-tail and erroneously bulrush, which strictly applies to *Scirpus* species.

A leaf and flower-head of reed mace *Typha domingensis*, used for mats and baskets.

4   Trees and Wooden Objects

It is a surprising fact that the land of Egypt was never short of timber for general construction in spite of the greater part of the surface area being desert. In pharaonic times, as now, the narrow Nile Valley supported trees of acacia, tamarisk, doum and date palms, Christ-thorn and sycomore. In the tropical areas of the south and in Nubia other timber species grow, especially ebony, which was brought from there to the pharaoh's palaces and temples.

A gaming board resting on ebony legs which are carved in the form of lions' feet standing on gold coils. The upper part has panels veneered with ebony and carved with heiroglyphics. The game pieces are missing – they were probably made of gold and stolen by the robbers.

However, it was to the countries north of Egypt that the pharaohs had to look for large timber and wood for special purposes, especially to Lebanon, where the cedar forests were regularly exploited by Egypt and neighbouring nations, such as Israel (1 Kings 5). Coniferous timbers such as cedar, juniper, fir, pine and cypress were the main species, but broad-leaf hard woods such as box, oak, storax and ash were imported from there and Asia Minor. Even the decorative bark of silver birch was included in Tutankhamun's tomb.

The skill of the Ancient Egyptian woodworkers was highly developed by the time of Tutankhamun. Consequently his tomb was embellished with an incredible array of finely crafted furniture that would have been

Photo: H Burton

admired by Chippendale or Louis XIV – some 3,000 years later. However, it must be said that not all of the objects were of equal quality. The practice of covering the surface of the wood with gesso – whiting and glue or gypsum – and then whitewashing, gilding and painting it obscured the imperfections of the basic construction. It also makes it difficult to identify what kind of wood has been used – one can hardly saw off pieces of timber from such precious items! Carter's initial descriptions 'white wood' and 'reddish wood', may apply to sycamore (p 58), and cypress (p 46) or juniper (p 60) respectively, but these are only guesses until microscopical identification is possible. Even so, much work has been done on available pieces by Boodle, Metcalfe and Chalk, and the following account provides notes on a selection of the wooden objects found in Tutankhamun's tomb followed by details of the tree species which provided the timbers. The glue used to fasten the pieces of wood together is mentioned in chapter 2 (p 21).

Finds from the Tomb

The golden shrines and coffins

Nobody was more surprised than Howard Carter when he looked for Tutankhamun's mummy in the burial chamber and found it in the centre of a *series* of boxes and coffins. The four large boxes are usually known as shrines, each smaller than the last and all covered with gold; the inner one contained the enormous stone sarcophagus in which lay the three golden coffins, with the pharaoh himself in the innermost one, which is made of solid gold and wonderfully ornamented. The outermost of the three coffins was made of cypress wood, according to Boodle, who studied two small pieces covered with gesso and thin gold foil. The coffin is shaped in the form of Osiris and is very heavy.

The framework of the gilded shrines is made of wood, and it was examined by A Lucas, the chemist attached to Howard Carter's team, whose detailed account of his findings is summarised here.

The outermost shrine measures 5 m (16 ft) long, 3.3 m (11 ft) wide and 2.8 m (9 ft) high, with the other three shrines smaller in order to fit into each other. Each section of each shrine was assembled in the chamber and the sections were fastened together with wooden dowels through holes in the planks, or by mortise and tenon joints.

It was possible for Lucas to study the planks themselves only where the ends joined the next section. In each case he concluded that the frame was composed of

Carpenters engaged in the various processes of primitive woodworking: *top left* smoothing timber, *top right* applying glue from a pot on the fire, *lower left* sawing, *lower right* adzing. From the tomb of Rekhmire.

cedar wood, but Boodle at Kew later identified a fragment which was still gilded as cypress (*Cupressus sempervirens* p 46). Of the 177 dowels examined by Lucas, 70 were of Christ-thorn (*Ziziphus spina-christi* p 68) and 107 of cedar wood. Other dowels were of copper, one of oak (*Quercus aegilops* p 48) according to Boodle and another of acacia wood (*Acacia* sp. p 22).

Thrones, chairs and stools

The great variety of seats was quite remarkable. They ranged from magnificent thrones to lowly stools, each having a timber framework. Nobody has made precise identification of all the species involved except for the well-known ones of cedar (*Cedrus libani* p 45) and ebony (*Dalbergia melanoxylon* p 46).

The royal throne (see pl 28) was covered with gold and decorated with glass, faience and stone inlay, so neither the identity nor the colour of the wood have been reported. Cobras and lions' heads were carved on it and its feet were also like those of a lion. Embossed in the gold overlay are wonderful scenes which Carter claimed to be 'the most beautiful thing that has yet been found in Egypt' – a claim that still holds true. He was referring in particular to the back panel facing the seat (see pl 4). In this scene Tutankhamun sits on a similar throne in his palace while his wife stands with one hand touching his shoulder. The whole picture is full of botanical interest, as well as having great beauty and significance. Both the figures, for example, are wearing full floral collars such as the one actually placed in his coffin (see pp 9–10). Moreover, on either side of the figures are huge floral bouquets (see p 9), one on one side and two on the other – composed of papyrus, lotus flowers and petals, mandrake

Photo: H Burton

Cedar wood chair back with an open-work panel.

fruits and poppy flowers. The outside back of the seat depicts a golden scene of a papyrus clump growing in water.

A lesser throne or armchair was of ebony and ivory, of a size suitable for a child – probably from his childhood home. Another child's chair was found in the Antechamber with a 'decorative panel of ebony, ivory and gold', while one in the Annexe was superbly made of wood decorated with ivory, gold and leather. This contrasted with what Carter described as a 'rushwork chair' (probably of reed), 'extraordinarily modern-looking in appearance and design'. The best carved chair of all is one of natural cedar wood with only a little gold embellishment – other gilded ornamentation below the seat was ripped out by the robbers, who did the same to the great golden throne.

The variety of stools is interesting too. Of note is the footstool with representations of bound captives on top which reminds me, as it did Carter, of King David's psalm: 'sit at my right hand until I make your enemies your footstool' (Psalm 110: 1, also quoted by Jesus in Matthew 22: 44 etc). Other sitting stools were four-legged, with simple or elaborate open-work side panels and painted white. One has a rushwork seat; another is of ebony and 'redwood', possibly cypress. The one I like is a delightful folding stool made of ebony with the ends of the cross legs carved in the form of elongated ducks' heads.

Beds and couches

It is difficult to distinguish between beds and couches so I shall refer to them all as the latter. They were certainly intended for sleeping on, as they were wide and designed to be lain on full-length, with a foot panel (but no head panel, as was the usual Egyptian custom). Each couch was carved with extraordinary representations of animals with elongated bodies and curious or realistic heads – lion, cow and composite hippopotamus-crocodile – and one had a huge arching tail, all gilded. The framework was made to be carried into the tomb in four pieces and reassembled, with the mattress fastened by bronze hooks onto staples. The mattress of one couch was of woven cord on an ebony framework; the cords would have been composed of fibres such as palm or papyrus rind (p 33), but the other timber used for the bedsteads has not yet been identified.

Caskets and boxes

Some of the most beautiful objects in the tomb were the ornate caskets, to which reference has already been made in several chapters. Their construction, however, was not always as perfect as their appearance. Carter commented that a certain casket inlaid with ivory and ebony as a marvellous marquetry of over 45,000 small pieces, was of inferior construction probably of tamarisk wood (*Tamarix aphylla* p 48). However, a reddish-brown coniferous timber, probably juniper (*Juniperus* sp. p 60) or cypress (*Cupressus sempervirens* p 46), was used for another casket, which had a remarkable lid with Tutankhamun's name on it in finely carved ebony and ivory hieroglyphs within a border of ebony, making a complete cartouche of the lid. Both of these caskets had been filled with jewellery, most of which was taken by the robbers, only pieces of less value were left, stuffed together in disorder (see p 42).

Chariots

The jumble of chariot wheels, sawn-up axles and decaying harness that Carter and his colleagues viewed with dismay were, in fact, magnificent state chariots such as had never been found before (see p 42).

They were constructed of wood and covered with sheet gold that was beautifully ornamented with figures of alien foes of the pharaoh, animal deities and floral designs (p 13). One of the chariot axles was actually decorated with birch bark. Fragments of timber examined by Boodle at Kew proved to be of elm (probably *Ulmus minor* p 49), which must have been imported from Western Asia. Whether the entire chariot was made of elm or a mixture of timbers is still uncertain. The remarkable point about its construction is that the grain of the curved timber was straight, so it must have been artificially bent, presumably by being steamed and pulled into form. Carter and later researchers examined in detail the techniques adopted by the coach-builders and concluded that, although the pieces were over 3,000 years old, one could hardly better them today. The dazzling effect of these gilded chariots drawn by plumed horses in full ceremonial parade must have been amazing.

Besides the state chariots, there were lighter ones (later known as curricles) for exercising purposes. Littauer and Crouwel have studied all the chariots in great detail and published their findings in volume 8 (1982) of the Tutankhamun Tomb series.

Model ships

Mention has already been made of the model papyrus skiff on which stood a golden statue of Tutankhamun (p 30, and pl 2). There were also numerous models of other craft, ranging from canoes to a fully rigged ship

Tutankhamun's ebony ecclesiastical throne with footstool. The back is inlaid with numerous pieces of faience, glass and stone, while the footstool, in gold, ebony and cedar wood, has representations of the king's foes, on which he would place his feet.

Photo: H Burton

Photo: H Burton

Wooden chests in the innermost Treasury with the gold-covered canopic canopy guarded by a golden figure. The head of the sacred cow, wrapped in linen, was made of wood and the horns and plinth painted with black varnish.

(pl 29). Unfortunately most of them had been broken by the tomb robbers, but they have been skilfully repaired and reconstructed. They were intended to enable the deceased to be ferried independently to his eternal rest – so that, as Carter put it, they 'follow the divine journeys of the sun . . . by day over the heavenly ocean, [and] by night [they] traverse the realms of Osiris'. Carter also explained that the models were made from 'logs of wood pinned together, shaped and planed with the adze'. He observed that they represent planking laid edge to edge, so that it presents a smooth surface on the outside 'fastened on the inside with tree-nails, having no ribs, but thwarts or cross-ties to yoke the sides, the side planking being fixed fore and aft to the stem and stern pieces. They have a steering-gear consisting of two large paddles which operate upon upright crutches and overhanging cross-beams before the poopdeck.'

Full-sized paddles or oars were also present in the tomb, placed between the outermost shrine and the north wall. There were ten of them, all pointing in the same direction, with the blades facing eastwards – I suppose the correct position for paddling to the west, since they were to ferry the king across to the Nether World.

Many of the models still retain a papyrus-craft shape, with curving stem and stern, and even have prows in the form of a papyrus inflorescence like that adopted for column capitals. The most impressive models are the four that are fully rigged and carry a square sail. Linen (p 34) would have been used for the sail, and for the rigging they used the fibres of flax, palms or papyrus rind (p 33). Such ships were required to tow the funerary barge, which was only equipped with steering oars.

In full-sized ships the tree trunks used for the masts would have been of the Cilician fir (*Abies cilicica*) from Lebanon. This tree was also traditionally used for the flag-poles that were fitted into the slots left at the sides of the doorway of each temple pylon. Other parts of Nile ships were normally made of local acacia wood, even to the masts, but royal ones were of cedar, like the Cheops boats buried in large pits beside the Great Pyramid (p 45).

Magnificent chariots jumbled together, and golden couches and wooden stools were all stored in the Annexe.

Wooden bows and reed arrows

We should not be surprised that a pharaoh was buried with his bows and arrows, since these were one of the Ancient Egyptian's principal weapons for defence and attack. What may surprise us, however, is their number, variety and ornamentation.

Altogether forty-seven bows were discovered in Tutankhamun's tomb, of which fourteen were simply cut from one piece of wood (technically called 'self bows'), while thirty-three were compound bows, that is they were composed of several pieces of wood glued together as a laminated stick and carved into a bow form. The strings, incidentally, are of animal gut, with only one of linen thread. Details of these bows have been published by W McLeod in volumes 3 (1970) and 4 (1982) of the Tutankhamun Tomb series.

The fourteen self bows were undecorated and gilded. For centuries this type was the normal weapon of soldier and peasant and Tutankhamun's were carved from a hard reddish wood. This may be acacia, which is likely to have been locally available, although any timber, local or imported, could have been used to ensure the qualities that a bow required. It is worth recalling that in England bows were traditionally made in this way from yew trees (*Taxus baccata*). No botanical examination of Tutankhamun's self bows has been carried out yet.

The thirty-three composite bows are very interesting from all points of view. This more powerful bow was invented in Western Asia towards the end of the Second Intermediate Period, about 1600 BC, and gained immense popularity until the time of Rameses III, about 1100 BC, when they seemed to go out of favour. There has been discussion as to whether Tutankhamun's bows were made in Egypt or in Asia, as the timber is of a foreign ash tree, either the manna ash (*Fraxinus ornus* p 47) or Syrian ash (*F. syriaca* p 47) with a covering of birch bark (*Betula pendula* p 45) which Carter described as 'cherry-like'. However, W McLeod considers that the importation of the timber and bark would have been possible and 'that there is no real reason to believe that any Egyptian composite bow was made outside Egypt'.

Egyptologists agree that the decoration of these bows is superior to any others. A protective layer of horn covered the wooden core with decorative bark outside the horn. Most of the ornamentation consisted of gilt or coloured geometric designs, hieroglyphs and the royal cartouche with various botanical motifs, including petal garlands, buds and clumps of papyrus.

The arrows were usually simple shafts and dozens of them were buried in Tutankhamun's tomb. They are made of reed (*Phragmites australis* p 35) fitted with metal or wooden heads. At the other end the feathers were inserted into a piece of wood that was spliced on to the reed. Usually in Ancient Egypt the wood used for this portion is local acacia or even imported box or ebony, but the only arrow examined by Boodle at Kew had a piece of almond wood inserted. Almond (*Prunus dulcis* p 62) was well known for its fruit, so it is not surprising that its timber was also used; there is in the Kew Museum a walking-stick from the 18th Dynasty made from almond wood.

Throw-sticks, batons and clubs

Throw-sticks – often called boomerangs in books – are frequently represented in Egyptian wall paintings of noblemen hunting wildfowl among the papyrus clumps. In Tutankhamun's tomb there were numerous hard-wood throw-sticks and even a pair made of an alloy of silver and gold called electrum. There were also weapon-sticks of different types. Several of these still had their natural bark on them, others were decorated with birch bark similar to some of the bows and one was covered with iridescent beetle-wings! Most sticks were offensive weapons with a knob at one end or more or less sickle-shaped clubs made of ebony (p 46), Christ-thorn (p 68) and tamarisk wood (p 48), ornamented with gold and ivory. Some of the sticks (or ceremonial batons) had been attacked by insects such as wood-boring beetles, shown by a fine powder among the linen cloths in which the sticks were wrapped.

TREE SPECIES

Abies cilicica
Cilician fir
Family: Pinaceae

Although the Cilician fir tree grows on Mount Lebanon, in Syria and Asia Minor, it was well known in Egypt on account of its coniferous resin, used in mummification processes (pp 21–2). It was known also for its tall straight trunks which were imported as ships' masts and as flag poles for the front of temple pylons. Loret thought that this tree produced the famous *ash*-wood (not to be confused with *Fraxinus*), but Meiggs considered cedar of

Two of Tutankhamun's ceremonial wooden sticks richly ornamented with gold and birch bark.

Photo: H Burton

Lebanon as the correct identification. A jar of resin labelled *ash* was found in Tutankhamun's tomb.

The Cilician fir is a stately tree growing up to 30 m (100 ft) with a straight little-branched trunk. The evergreen needle leaves are spread equally along the branchlets. Its cylindrical cones are about 20 cm (8 in) long and reddish-brown with large flat scales.

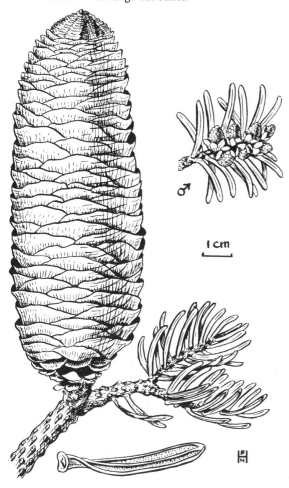

A cone of the Cilician fir *Abies cilicica* said to yield the *ash* of Ancient Egypt.

Betula pendula
Silver birch
Family: Betulaceae   (see pl 30)

It is very easy to understand why the attractive silvery bark of the birch tree was used to decorate items such as bows, a bow box, sticks, a fan handle, goads and a

chariot-axle in Tutankhamun's tomb, but it is more difficult to envisage the transport of the bark to Egypt. As a non-Egyptian tree it would have been necessary to import it from central Turkey or even mountainous areas of southern Europe.

Further north the silver birch is a common lowland deciduous tree up to 25 m (80 ft) high with a graceful habit and thin peeling bark. The oval or diamond-shaped leaves are toothed. Its male flowers are borne in hanging catkins, the female in less conspicuous upright ones.

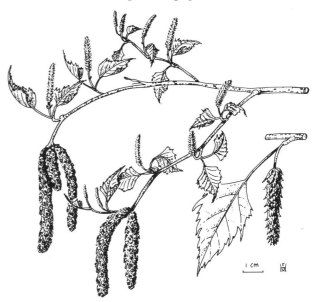

Male and female catkins of silver birch *Betula pendula*, and a fruiting catkin with mature leaf.

Cedrus libani
Cedar of Lebanon
Family: Pinaceae

This famous coniferous tree grows on Mount Lebanon from where it had been exported to Egypt since early times – even in the Old Kingdom, 4th Dynasty it was used for Cheops' ship which was found at the base of the Great Pyramid at Giza. Cedar timber is reddish and fragrant and it was also used in Solomon's temple at Jerusalem (1 Kings 5–7) and in Nebuchadnezzar's palace at Babylon.

The tree is up to 40 m (130 ft) high, with spreading horizontal branches in old individuals and a stout trunk up to 2 m (6 ft) or more in diameter. The evergreen needle leaves form in clusters and the large cones stand erect until they are mature when they fall and shatter, releasing the seeds.

A cedar of Lebanon forest. *Cedrus lebani* was greatly used in Ancient Egypt for ships and furniture.

Cupressus sempervirens
Cypress
Family: Cupressaceae

Splendid reddish fragrant cypress timber was used for the outer coffin. It must have been obtained from outside Egypt, perhaps from the mountains of Edom or further north, from Syria or Turkey, from where it could have been transported by sea.

This is an evergreen tree growing up to 30 m (100 ft) high forming a trunk up to a metre across. Its mass of dense branches may be columnar (in cultivated forms) or may spread horizontally (as in the typical wild form). The twigs are covered with small scale leaves. Its cones are 2–3 cm (about 1 in) across and almost round, with grooves which open to release the winged seeds.

Its oil is obtained from the leafy shoots, rather than as resin from the wood, as in other conifers. The wood is known to have been used for several other Egyptian coffins.

Dalbergia melanoxylon
Ebony
Family: Leguminosae

It is interesting to note that our English world 'ebony' is derived directly from the Ancient Egyptian hieroglyph *hbny* which refers to a dark, if not black, timber. This hard, expensive wood was used for special furniture such as Tutankhamun's chair, stool and bed. Pieces were also incorporated in the shrine doors as bolts, and were used for a gaming-board stand, legs of another chair and the framework of boxes. Even part of the doorframe into the burial chamber was of ebony (combined with pine and date palm). Thin sheets were used for inlay and as a veneer of high technical standard. However, the black guardian statues were painted, presumably to look like ebony, as were several other objects in Tutankhamun's tomb.

Botanically the Ancient Egyptian 'ebony' is a different timber from the tree in the genus *Diospyros* now known by that name. The dark wood found in Tutankhamun's tomb and others came from *Dalbergia melanoxylon*, a member of the pea family. It occurs in dry wooded grassland south of the Sahara, from where it would have been imported to Egypt. It grows as a spiny shrub or tree 5–30 m (15–92 ft) high, often with several trunks, none of which are very thick. The heart-wood is purplish brown,

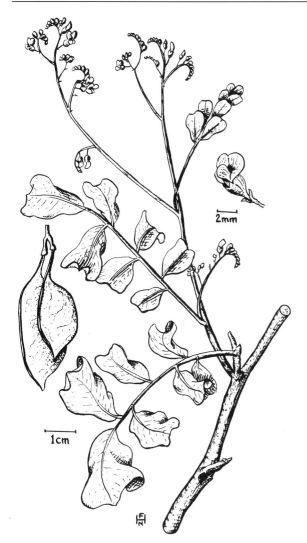

A branch from the ebony tree *Dalbergia melanoxylon*.

verging to black. The leaves are compound with three to four pairs of small leaflets, and its pea flowers are small, white and fragrant. Since the timber is immensely hard, one wonders how the Ancient Egyptians managed to work it and in particular, how they made sheets of veneer from it. As the Egyptians were seeking a jet-black timber, the name 'ebony' later became attached to *Diospyros* species from tropical Africa, which have almost black wood and, even later, to *Diospyros ebenum*, from India and Sri Lanka, which is now the main timber known by that name.

Fraxinus species
Ash
Family: Oleaceae (see pl 31)

Ash wood was used to make the composite bows (p 43) found in the tomb. One specimen was identified by Chalk as '*Fraxinus* species probably not *F. ornus*' while another fragment examined by Boodle was positively named as *F. ornus*, the manna ash.

The manna ash is a small, rounded tree occurring in woodland and maquis-type vegetation from Lebanon to Italy. It carries pinnate leaves and conspicuous inflorescences of white flowers in springtime. The fruits are typical ash keys, which rotate as they are blown from the tree.

Apart from the manna ash, there is the Syrian ash (*F. syriaca*), which occurs in the eastern Mediterranean area and could have been imported. This is a taller tree, up to 15 m (50 ft) high, with greenish-brown flowers in smaller clustered inflorescences appearing before the leaves. Other possible species could be *F. angustifolia* and *F. pallisiae*. Ash timber is strong and resilient and so would have been ideal for bows, as well as for cartwheels and tool handles. (It should not be confused with the heiroglyphic word *ash*-wood p 20).

Liquidambar orientalis
Levant storax
Family: Hamamelidaceae

This tree is normally included among resiniferous species instead of timber trees. In this case it is here because a piece of its wood was found on the floor of Tutankhamun's tomb, about 18 cm (7 in) long with almost square section (8 × 10 mm). 'One end is shaped like the cutting

A branch from the Levant storax *Liquidambar orientalis*.

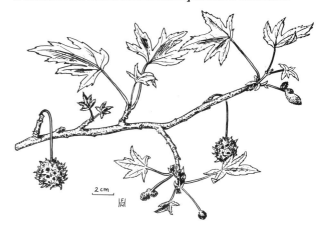

end of a chisel and the other end is square. In the tomb records there is not any mention of this . . .'. Dr Metcalf identified it at Kew.

The tree has a rounded habit, with drooping branches and palmately divided alternate leaves. Its inflorescences are spherical, suspended on stalks several centimetres long. It occurs only on Rhodes and in south-east Turkey, where it has long been tapped for its resin, usually called balm, Levant storax or styrax, but there is much confusion both over the use of these names and over the various species which yield the resin. This is obtained by beating the bark in spring, which causes the resin to accumulate under the bark until it is collected in the autumn. After it has been extracted with boiling water and has been purified the viscid greyish liquid turns yellow-brown. It was used in Egyptian perfumes and for mummification (pp 20–2).

Quercus aegilops
Valonia oak
Family: Fagaceae

When Boodle was sent some pieces of wood from the cross-tongues or tenons of the second, third and fourth shrines he at first identified them as Turkey oak (*Q. cerris*). Later he concluded from further examination that they were Valonia oak (*Q. aegilops*). The latter name includes a wide range of variations in the eastern Mediterranean oaks, such as *Q. ithaburensis*, which are difficult to classify.

The Valonia oak has a distribution from Palestine to Turkey and Greece, occurring singly or in woodland groups. It is a fair-sized deciduous tree with a stout trunk yielding good strong timber of a light colour similar to other common oaks of Europe. The oblong leaves are about 8 cm (4 in) long and are sharply toothed. The catkins are of separate male and female flowers, and the acorn fruit has a cup covered with long recurved scales.

Tamarix aphylla
Tamarisk
Family: Tamaricaceae   (see pl 32)

In Eygpt there are several native species of tamarisk which superficially appear to be similar. They form shrubs or small trees with numerous twigs, giving the whole plant a feathery appearance. On closer inspection it will be noticed that the twigs are covered with small green scales which are reduced leaves often covered by secreted salt. *T. aphylla*, however has no obvious scales. The small pink

Photo: F N Hepper

Corky bark of an old tamarisk tree *Tamarix aphylla*.

or white flowers are held in short terminal and lateral racemes like little bottle brushes.

Tamarisks are plants characteristic of salty places, riversides or dried-up stream beds (wadis) where water flows from time to time after rain. They are planted in such situations, where few other trees grow, in order to form hedges or wind-breaks; they are easily propagated from short, woody cuttings. Deep roots develop as well as side roots, which effectively hold sand; *T. aphylla* stores a large amount of water in its roots. The shrubby species hardly develop a trunk, while the trunk of the tree tamarisk (*T. aphylla*) is the most likely one to have been

A shoot of kermes oak *Quercus aegilops*, which provided the timber for some of the tenons in Tutankhamun's burial shrine.

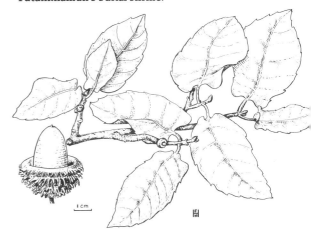

1 cm

Plate 25 (*top left*) Flax plants *Linum usitatissimum* in flower at Kew.

Plate 26 (*above*) An ornate cabinet with a framework of ebony and panels probably of cedar of Lebanon.

Plate 27 The paintings on this beautiful casket found in the Antechamber depict dramatic hunting and battle scenes. Local desert flora are depicted, and the scenes are bordered by mayweed flowers.

Plate 28　Tutankhamun's golden throne. Sheet gold overlays the wooden frame. Lions and winged serpents are conspicuous ornaments, with botanical motifs represented by a papyrus swamp on the back panel and the floral collars and bouquets on the front panel. See also pl 4.

Photo: Robert Harding Picture Library

Photo: F N Hepper

Plate 29 A model ship with full rigging of the type that would have been used to pull the funeral barges. Royal ships were built of cedar and were equipped with sails and ropes of papyrus rind or palm fibres.

Plate 30 Bark of the silver birch *Betula pendula*, such as encrusted Tutankhamun's bows and chariots.

Photo: F N Hepper

Plate 31 Flowers of the manna ash *Fraxinus ornus* growing at Kew.

Plate 32 A salt-encrusted tamarisk tree *Tamarix aphylla* at Kharga Oasis.

Plate 33 A leafy shoot of the elm tree *Ulmus minor*, the timber of which was used in Tutankhamun's chariots.

Plate 34 Water-melon fruits *Citrullus lanatus*.

Plate 35 Cocculus *Cocculus pendulus*, with an acacia tree *A. tortilis* in the background, in Wadi Araba.

used for timber in Ancient Egypt: it is dense and useful for general construction, turnery and as fuel. A sweet substance, said to be the manna collected by the Israelites (Exodus 16), is secreted by insects sucking juice from the twigs of some species.

Ulmus minor
Elm
Family: Ulmaceae   (see pl 33)

Elm wood was used for Tutankhamun's chariots (p 40) and it was artificially bent during construction. The timber must have been imported, since elm trees are not native to Egypt.

Boodle at Kew identified the timber with *U. nitens*, a name not now recognised, being a synonym of *U. minor*, the European field elm or smooth elm. It has a wide distribution from Britain to Turkey and Algeria, although much reduced in recent times by Dutch elm disease. The classification and nomenclature of elms is exceedingly complicated and the identification of timber to species level is uncertain.

The field elm develops into a tree 20 m (60 ft) or up to 30 m (100 ft) high with a good trunk. Its timber is brown, and of medium weight, with a rather twisted grain which may make it difficult to work – even so, in Europe, it has long been popular for many purposes including wheel-making.

*   *   *

For other timber trees, see: acacia (*Acacia* sp. p 22), sycomore (*Ficus sycomorus* p 58), juniper (*Juniperus* sp. p 60), olive (*Olea europaea* p 16), pine (*Pinus* sp. p 26), almond (*Prunus dulcis* p 62), willow (*Salix subserrata* p 17) and Christ-thorn (*Ziziphus spina-christi* p 68), see also the reed (*Phragmites australis* p 35).

5   Food and Drink

As it was considered essential to provide the *ka*, or soul of the deceased, with food similar to that consumed during life, graves of the wealthy were well stocked with all manner of foodstuffs. The dry atmosphere preserved those in Tutankhamun's tomb in a remarkably good state.

Finds from the Tomb

Fruits, nuts and seeds

A very good representation of the fruits and nuts available during the 18th Dynasty was included in Tutankhamun's tomb. There were samples of water-melon (p 56), syco-more fig (p 58), grewia (p 59), doum palm (p 59), persea (p 15), almonds (p 62), dates (p 62), pomegranates (p 62), grapes (p 67) and Christ-thorn (p 68). Others that might have been expected were common fig (*Ficus carica*), argun palm (*Medemia argun*), olive (p 16) and cordia (*Cordia gharaf*). It is as well to mention that some fruits that have been reported from Egypt were in fact latecomers belonging to Graeco-Roman times. Such species include carob (*Ceratonia siliqua*), various species of *Citrus*, peach (*Prunus persica*) and apricot (*Prunus armeniaca*). Wine was made from grapes although other fruits such as dates were also used, and honey certainly formed the basis of a mead.

Honey

Honey is made by bees from nectar collected from innumerable flowers. Bees were kept in clay tubular hives by professional keepers who had to put up with their stings. Honey was an important sweetener for the wealthy of Ancient Egypt long before sugar came into general use, and jars were included in some tombs. In Tutankhamun's tomb, one was actually labelled 'honey of good quality'. A sample of honey from a 19th Dynasty tomb was analysed by E Zander, who found the pollen consisted

Grape harvesting for wine-making in Ancient Egypt. Copy of an illustration from the tomb of Nakht (18th Dynasty).

principally of persea (*Mimusops laurifolia* p 15) and Egyptian plum (*Balanites aegyptiaca* p 23) with a few grains of grasses, legumes and other wild herbaceous plants.

Wine and wine-jars

Vines are planted either individually for grapes for home consumption or more usually in vineyards for the production of fruit in quantity and for wine. Water during the winter is important and in the rainfall areas of the Mediterranean regions terraces of vines are a common sight. In Egypt irrigation must provide moisture. The roots of the vine are extremely well developed and they can penetrate fissures of rock deep below the surface.

Pruning takes place during the period of winter dormancy, and the previous year's growth is cut back to one or two large buds at the junction with an older stem. This limits the number of shoots and allows the roots to nourish fewer clusters of grapes, of better quality.

Close to the vineyard there would have been vats in which the grapes were collected and trodden by workers. As the juice ran out it was collected into further vats for fermentation and the residue squeezed out in cloth bags. Grape juice or 'must' ferments naturally since it contains a minute yeast organism (*Saccharomyces cerevisiae* var.

decades earlier. Four jars were labelled 'sweet', but no doubt all of them were choice wines, fit for the royal table.

Such information is found on amphorae throughout Ancient Egypt and it was not unique to this tomb. There is a famous mural in one at Thebes – that of Nakht who lived during the reign of Sethos I in the 19th Dynasty – in which the vine-dressers are shown picking grapes from an arbour, while other men are pressing out the juice with their feet. A row of wine-jars depicts the finished product and it is possible to see the impressed dockets on the side of the dried mud plugs similar to those actually found in Tutankhamun's tomb.

Of course, the wine itself had disappeared in the course of time, but sometimes in such jars its residue remained, together with yeast and even pips; two jar residues analysed by Lucas revealed substances consistent with wine. Carter assumed that the jars were coated with a resin in order to render the pottery less porous to the liquid, but Lucas examined twenty-two broken jars and found no evidence of resin coating.

After being filled with fresh wine the jars were closed with a bung of vine leaves, chopped straw or rushes, and sealed with mud. Carter explained how it was possible to see the small hole in each plug that was left to allow residual carbon dioxide to escape. Only when this fermentation was complete were these holes filled and the wet mud stamped with a small seal to indicate the vintage. Such information was written in the hieratic script, which was the popular version of the Egyptian hieroglyphs. Wine features in many inscriptions and the Egyptians were regarded by classical writers as wine-bibbers. Wine was also used medicinally and as an antiseptic for wounds. Almost any fruit, such as dates and pomegranates, as well as grapes, could be used for wine-making.

Vegetables, herbs and spices

In this book the distinction between fruits, vegetables, herbs and spices is made on a traditional basis, since botanically there is a great deal of overlap. For example, garlic is classified as a vegetable, together with onions and leeks, although garlic could also be considered to be a flavouring spice. Chickpeas and lentils are seeds, but also reckoned to be vegetables, as are other legumes known from tombs in Ancient Egypt – broad beans (*Vicia faba*) and peas (*Pisum sativum*). Salads and vegetables popular during Tutankhamun's lifetime included cos-lettuce (*Lactuca sativa*), radish (*Raphanus sativus*), cucumber (*Cucumis melo*) and chicory (*Cichorium pumilum*).

The herbs and spices in the tomb are representative

392   571

Wine jars in Tutankhamun's tomb showing the mud seals impressed with details of the vintage.

ellipsoideus), which breaks down the sugar into alcohol and carbon dioxide. After six weeks' fermentation the new wine is left for another month while the sediment, or lees, of pips, skins, stalks and yeast settles to the bottom, then the clean wine is decanted off it into jars.

Three dozen pottery wine-jars were found, labelled with the place, date and vintage of the wine, in the same way that French wines of today bear their *appellation*. Howard Carter reported that Tutankhamun's wines were prepared in the Delta of the Nile, which fits in well with what is known about vine cultivation in Egypt, as they flourish there. In 1965 Černý published a more detailed translation of the labels, showing that most of Tutankhamun's wine came from an area called the 'House of Aton' in the Western Delta, with one jar from the north-east, and another possibly from Amarna. However, one jar was from the 'Fruit of the Southern Oasis', which comprised Kharga and Dakhla. The name of the chief vintner was stated, as well as the date of Tutankhamun's reign as the fourth, fifth, ninth and tenth year, while one was dated 'year 31', referring to the reign of Amenophis III, several of those known to have been used during the 18th Dynasty. Coriander seeds (p 57), juniper berries (p 60), black cumin seeds (p 61), a sprig of wild thyme (p 64) and fenugreek seeds (p 64) were found. To these would have been added cumin (*Cuminum cyminum*), dill (*Anethum graveolens*), and various mints (*Mentha* sp.), as popular herbs at that time.

Cereals

The great civilisations of the Near East were dependent on cereals as their staple food. Both wheat and barley – 'the corn of Egypt' – originated in hills of the Fertile Crescent from Syria and Turkey to Iran, where the primitive ancestors of our modern cereals grew in clearings in the oak woodlands.

The important difference between the wild and cultivated plants is that while the ears of the wild ones shatter and break up on ripening, the cultivated plants remain intact. This was a very important development, as it would have been impossible to cut and carry sheaves of wild cereals to a storehouse without losing the grain on the way. A crop of cultivated cereal with non-shattering ears will not disintegrate while being cut and stacked. On the other hand threshing became necessary to separate the grains and the cereals actually became dependent upon man for their continuation. It is interesting to note

Harvesting cereals and storing grain in a granary in Ancient Egypt. From J G Wilkinson *Manners and Customs of the Ancient Egyptians*, London, 1878.

how careful Joseph was to safeguard sowing grain, as once eaten it could not be restored (Genesis 47:23–4). A later development in cultivated wheat enabled the grain to lose its outer husks (to form 'naked' wheat) instead of having them closely covered by chaffy scales (as in 'hulled' wheat). No doubt our ancestors gathered wild wheat long before they cultivated it. In a good year in the right place they would have been able to gather ample supplies for each family, but they would have had to have done it before the ears shattered and the birds devoured the standing and fallen grain. Agriculture became established as people harvested and sowed the grains which developed non-shattering ears, while the stored grain and farm plots necessitated an increasingly non-nomadic mode of life, instead of a hunter-gatherer type of existence. The botanical history of cereals, described by Daniel Zohary and Maria Hopf in *The Domestication of Plants in the Old World* (see Plant Species References) is an interesting and complicated story.

The civilisation of Ancient Egypt relied heavily on cereal cultivation. Ideal conditions were available in the flood-plains of the Nile Valley, making the corn of Egypt proverbial to surrounding nations such as Israel, especially in times of famine. But Egypt herself was not immune to famine, as shown by the story of Joseph in Genesis 41 onwards. Since no rain falls in Egypt (or at least not enough for the cultivation of crops) agriculture depends on the River Nile for water. Even the Nile fluctuates, since it is governed by the rainfall in tropical Africa and Ethiopia, as we know from severe droughts during the 1970s and 1980s. However, nowadays the dam at Aswan

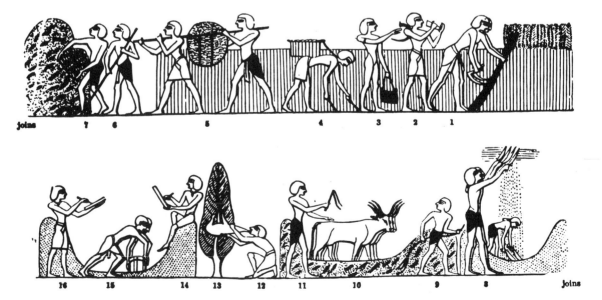

joins   7   6   5   4   3   2   1

16   15   14   13   12   11   10   9   8   Joins

impounds water in Lake Nasser and releases it gradually into the Nile for irrigation purposes, saving Egypt from famine in recent times. Unfortunately this means that the fertile silt is no longer deposited on the fields nor is accumulated salt washed away. Over 2,000 years ago Herodotus called Egypt 'the gift of the Nile', but without the unremitting organised labour of its inhabitants the full benefit of their gift could not be enjoyed.

A model granary

Among the items classed as funerary equipment is a delightful model of a granary. It is almost square, 74 × 65 cm (29 × 26 in) and 21.5 cm (8½ in) high with a doorway entrance and internally divided into sixteen compartments. These were for storing the grain, and indeed this model contained real seeds 'up to the brim', according to Carter's account. Although the model is of wood, the actual granaries would have been constructed of sun-dried mud brick.

This model housed emmer wheat for the most part, with seeds of fenugreek, chickpea and various weed seed contaminants. Often in Ancient Egyptian tombs one finds such seeds, which help to provide a picture of the ecology at the time of harvesting. How marvellous to be able to

Photo: H Burton

A model granary included in the tomb.

visualise the wild clovers, grasses and thistles inhabiting a cornfield thousands of years ago! A sample of the mixed grains was sent by Carter to Boodle at Kew in 1933 and he forwarded the wheat to an expert on cereals, Professor J Percival, who identified it as emmer. There are spikelets

The loaves found in Tutankhamun's tomb were enclosed in a network of rushes. They mostly contained fruits of Christ-thorn as well as cereal flour.

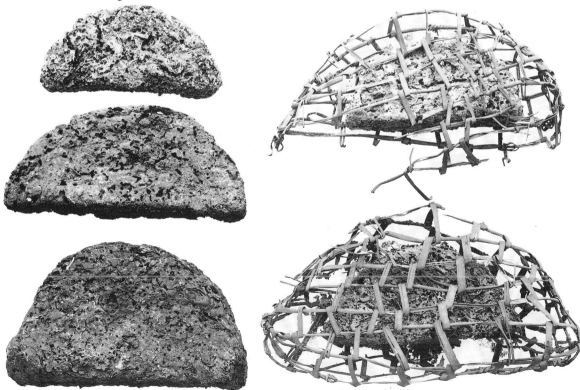

Photo: H Burton

of mature grains, some spikelets being entire, others broken (see p 66).

Bread

A dozen or more loaves were deposited in Tutankhamun's tomb and scattered by the robbers in their rush for gold. Most of them were semi-circular in shape, and they varied in size up to 13 × 7 cm (5 × 3 in) and 2.5 to 5 cm (1–2 in) thick. Some were enclosed in a network made of rushes, as can be seen in the photograph on p 53, perhaps used as a carrier. One loaf was triangular with sides of 20 cm (8 in) in length and 3.5 cm (1½ in) thick. Since all of them contained fruits of Christ-thorn they were really fruit loaves or cakes. One lump seemed to be of plain bread, however.

Bread left in other tombs has often been found to contain considerable quantities of grit, probably from desert sand blown in as well as from the grinding-stones. Hence it is not surprising that the teeth of mummies are often found to be badly worn down, from eating so much grit. A model grinding-block (called a saddle-stone) with a rounded rock for crushing the grain, was included in Tutankhamun's tomb.

An Osiris bed

An oblong box 202 cm (80 in) long and 88 cm (35 in) wide was found in the tomb under chests and model ships. It had a lid which was slightly rounded on the outside and pegged on to the box. The whole object was varnished with black resin both inside and out.

On opening this box Carter found a flat model figure wrapped in linen. The bindings were carefully done and beneath several series of linen cloths he found a rough wooden framework in the form of the deity Osiris. Inside the hollowed form was a filling of clear quartz sand, evidently from the bed of the Nile, in which barley grains had germinated. The seedlings had grown about 8 cm (3 in) before they withered.

What was the purpose of this curious object? The clue is the figure of the cult-god Osiris who was worshipped throughout Egypt, but especially at Abydos and Busiris. The growth of the barley symbolised new life after death – one of the main themes of the Osiris cult. Plays depicting his supposed life and death were frequently performed, as was the ritual burial of germinated barley on a little bed in the form of Osiris' body – nowadays known as an 'Osiris bed'. Such Osiris beds continued to be placed in the tombs of other important people long after Tutankhamun's time.

Photo: H Burton

An 'Osiris bed' found in Tutankhamun's tomb with cereal grains which had germinated in the silt filling the wooden frame.

Beer

The presence of two beer-strainers among Tutankhamun's possessions is interesting, although as far as I know no jars of beer were found. They were made of wood, with a piece of punctured copper covering the hole. The purpose of the strainers was to filter off the chaff from the liquid as it was poured into another receptacle. In contrast, it might be mentioned that the Philistines of Lachish, a little later than the period we are talking about, drank beer from the short spout of a vessel having a filter incorporated in it, rather like a modern teapot.

Scenes of beer-brewing have been preserved in other tombs, showing the processes of preparation, fermentation and filtering. For fermentation to take place, the starch in the barley grains has to be changed into sugar – a process which occurs naturally in germination. When the grains begin to sprout they are dried to stop growth and then soaked in water with yeast (or dough containing yeast), which converts the sugar into alcohol. Filtering takes place when the process has gone on long enough for the brew to be ready to drink.

PLANT SPECIES

Allium sativum
Garlic
Family: Liliaceae

Some splendid bulbs of garlic with well-preserved leaves were found in Tutankhamun's tomb, although Newberry called them onions. Many other Egyptian tombs contained them too, left as offerings or associated with the mummy in armpits or groin. Garlic was a favourite food, as it is today, and it was used medicinally and as a preservative – which may explain its association with mummies!

Unlike the common onion, which has a bulb of concentric fleshy leaves, garlic has a number of succulent scales (known as cloves) joined at the base. These can be separated and planted to form new bulbs, as no seeds develop from the flowers. The inflorescence develops at the top of a stem about 50 cm (20 in) high and taller than the hollow leaves, but the greenish-white flowers do not set seed as one would expect. Such sterile behaviour indicates a plant of long cultivation and uncertain ancestry. In fact, botanists do not know of a direct wild ancestor and it is thought that garlic developed from species in Iran (which was ancient Sumeria).

Photo: H Burton

Well-preserved bulbs of garlic *Allium sativum* from Tutankhamun's tomb, now in Cairo Museum.

The Ancient Egyptians were also fond of onions (*A. cepa*) and leeks (*A. porrum*), hence the cry of the Israelites as they left Egypt for the wilderness of Sinai that they longed for these succulent items of food (Numbers 11: 5). Herodotus recorded the vast qualities of onions, garlic and radish consumed by the builders of the pyramids at Giza. Some Egyptologists now believe, however, that this was a misreading of inscriptions and that the vegetables were offerings rather than ration lists.

Cicer arietinum
Chick-pea
Family: Leguminosae

Only a few seeds of chick-pea were found, mixed with cereals and other grains in the model granary. This is one of the few times it has been found in an Ancient Egyptian

A plant of chick-pea *Cicer arietinum.*

site, although texts often mention it from the 18th Dynasty onwards.

This legume is an erect annual about 20 cm (8 in) high with pinnate leaves bearing up to eight toothed leaflets. The very small bluish or white pea flowers produce a short one- or two-seeded pod. Characteristically shaped seeds have given rise to various names, such as chick-pea, falcon-head and ram's-head (hence the scientific epithet *arietinum*, from the Latin *aries*, a ram) and it takes little imagination to see these creatures in the seed. A delicious paste called *hummus* is made from crushed chick-peas

mixed with sesame and olive oil and eaten with pitta bread.

Citrullus lanatus (synonym *C. vulgaris*)
Water-melon
Family: Cucurbitaceae   (see pl 34)

A large quantity of seeds of the water-melon were deposited in Tutankhamun's tomb in two oval baskets. They were still in remarkably good condition as flat seeds about 8 mm (5/16 in) long. When fresh they would have been lightly roasted and then eaten by cracking the brittle outer cover with the front teeth.

Water-melon seeds are as well known today in Egypt as they were in pharaonic times. They occur scattered in the usually red (or white) flesh of the large round dark green fruits, which are much bigger than a football. I find the taste of water-melon rather diluted as it is so watery – a feature acceptable, however, in the dry heat of a North African summer and which would have been welcomed by the wandering Israelites as they left Egypt for the desert of Sinai (Numbers 11: 5). The very long stems trail along the ground bearing yellow flowers among the hairy lobed leaves. The plant is an annual.

Botanists have concluded recently that the water-melon was developed as a cultivated plant in early African agricultural systems from its perennial tropical ancestor, the colocynth (*Citrullus colocynthis*). This is still found in Egyptian deserts, with clusters of yellow fruits about the size and shape of tennis balls, but it is a violent purge with dangerous seeds and pulp. Two other gourds occurred in Ancient Egyptian tombs, though not in Tutankhamun's. They are the cucumber (*Cucumis melo*) which is usually short and curved with numerous longitudinal grooves and used as a vegetable; and the bottle gourd (*Lagenaria siceraria*), a tropical species with bottle-shaped fruits used as containers when hollow and dry.

Cocculus hirsutus
Cocculus
Family: Menispermaceae   (see pl 35)

Mixed with the tropical grewia (p 59) were fruits of another tropical plant known as cocculus. It was identified by Schiemann of Berlin but I have not seen any in the sample of grewia at Kew. Whether it came there in error (the fruits of both species look superficially alike) or by intention is uncertain. It is known that a native Egyptian species of cocculus (*C. pendulus*) is used in medicine as a diuretic and the leaves are edible. Both cocculus species are wiry twiners or scramblers which creep along the dry ground and both have minute flowers and small round black fruits with a sculptured stone inside.

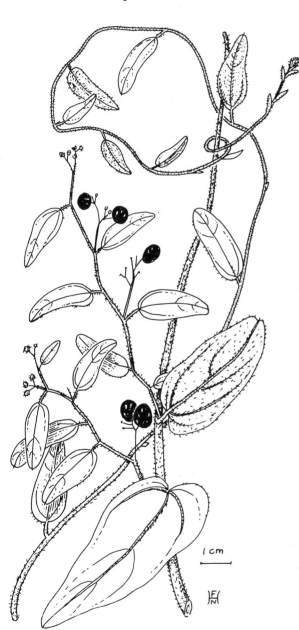

Shoots of cocculus *Cocculus hirsutus.*

Coriander *Coriandrum sativum.*

Coriandrum sativum
Coriander
Family: Umbelliferae

Several samples of coriander seeds were placed in Tutankhamun's tomb and they are still in excellent condition, though some of them have holes made by weevils. The seeds are enclosed within the hard fruit covering and they were harvested as one. These ancient samples are rather smaller than the average modern fruit size, but since there is a wide range in size even today we cannot be certain that their whole crop was smaller.

Coriander is an anciently cultivated crop plant allied to parsley and caraway. Its spicy seeds are used for seasoning meats, and they also yield the volatile oil of coriander which is used for flavouring purposes and has been used medicinally since ancient times. Coriander must have been well known to the Israelites who made their exodus from Egypt some 200 years after Tutankhamun, because they likened the appearance of manna in

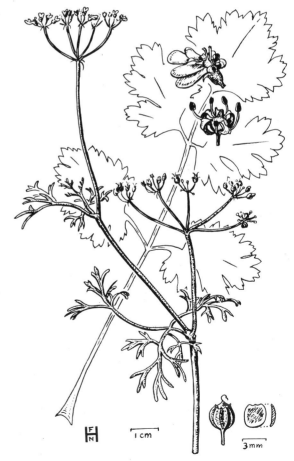

the desert to 'coriander seed, white' (Exodus 16: 31). The first records from Egypt date, in fact, from the 18th Dynasty, so it may still have been something of a rarity during Tutankhamun's reign, having originated in southern Europe. Coriander is now grown in warm countries throughout the world, especially in India.

The coriander plant is a slender annual growing to about 50 cm (20 in) high, with broadly lobed basal leaves which are eaten in soups and salads; the upper leaves are very different in appearance, having finely divided segments. The flowers are white, tinged pink, and clustered in small umbrella-like heads. Its round fragrant fruits are ribbed and about 2 or 3 mm in diameter.

Ficus sycomorus
Sycomore fig
Family: Moraceae   (see pl 36)

A medium-sized oval basket contained some sycomore fruits, which are smaller than those of the common fig (*Ficus carica*). The latter was known during the 18th Dynasty onwards and figs have been found in other tombs, but surprisingly they are not reported from Tutankhamun's.

The sycomore is a tree of warm climates and extends deep into tropical Africa. It is known in Egypt from at least the 1st Dynasty onwards and used to be a common tree to such an extent that the Psalmist recalls the disastrous effect of cold when God 'destroyed [the Egyptians'] vines with hail and their sycomores with frost' (Psalm 78: 47). Trees were planted near habitations where the fruit was readily available and advantage could be taken of their shade. The heart-shaped leaves are rough to touch and they persist throughout the year, except in cooler regions, where most of them may fall in winter.

The fruits develop all over the trunk and branches in large clusters. Technically each is called a syconium and it has an internal cavity lined by tiny florets and with a hole at the top of the fruit. In through the hole squeezes the pollinating wasp, which crawls around inside laying eggs in some of the florets and pollinating others at the same time. After some five weeks the young wasps hatch, the wingless males first, followed by the winged females, with whom they mate before the latter leave the cavity via holes eaten in the top of the fruit by the males. The whole life-cycle of the fig is remarkably adapted to the insect as the fruits ripen at the same time as the insects hatch. Unfortunately the presence of the dead male wasps makes them inedible, so to make them palatable sycomore growers in ancient times used to clamber over the trees

A branch of the fig tree *Ficus sycomorus*.

5 mm

1 cm

cutting each young fruit. It was once thought that this was done to liberate the insects, as the cut fruit was then without dead ones inside. However, recent research has shown that the cutting of the exterior causes the syconium to develop in a few days instead of weeks, thereby short-circuiting the insect's life-cycle. In fact, the full story is a complicated one, with several species of wasp being involved. Modern varieties of sycomore fig, however, are self-fertilised.

Egyptian sycomore fig trees live to a great age and become massive in size. Often they were cut for timber and their soft wood was greatly used for the construction of coffins (sarcophagi) and general woodwork in Ancient Egypt. So far none of its wood has been identified from the tomb of Tutankhamun. Note that the sycomore fig should not be confused with sycamore, the name which is applied to species of *Acer* and *Platanus*.

Grewia tenax (synonym *G. populifolia*)
Grewia
Family: Tiliaceae

It is rather surprising to find in several baskets fruits and seeds of a tropical shrub botanically known as *Grewia tenax*. (It has no English name.) Although grewia does occur in Egypt proper, its distribution is in the hotter southern parts about Aswan and the Red Sea coasts, far from the Valley of the Kings.

The shrub grows up to 3 m (10 ft) high and bears small toothed leaves, white flowers, and orange-red fruits. In tropical Africa, fruits of grewia are eaten by the local people, and we now know that some species are rich in sugar, citric acid and vitamin C. It is possible, therefore, that such an exotic species was considered a worthy delicacy for the pharaoh's tomb.

A leafy, flowering shoot of grewia *Grewia tenax*.

Hordeum vulgare
Barley
Family: Gramineae

Barley is an annual grass, usually smaller than wheat, with a shorter growing season and tolerating poor, dry soil. Its ears are always bearded (awned), with the grains arranged in six rows, with three on each side of the ear – hence its name 'six-rowed barley'. The primitive cultivated barley of earlier times had only two rows of grains, with one on each side and this 'two-rowed barley', scientifically called *Hordeum distichon*, has been found in archaeological excavations of ancient deposits. Cultivated barley does not shatter or break up, while its wild ancestor *H. spontaneum* scatters its grains in open places among the oaks on the hills of the Fertile Crescent.

Barley was more important than wheat – texts regularly refer to 'barley and wheat' in that order of priority, as cereal for bread. It was also popular as the principal cereal for brewing beer (p 55). Hail spoilt the growing barley during the plagues of Egypt (Exodus 9: 13–35).

Barley *Hordeum vulgare*.

Hyphaene thebaica
Doum palm
Family: Palmae   (see pl 37)

In the tomb there were several baskets containing well-preserved doum (also spelt dom or dum) palm fruits the size of tennis balls but more or less cubic in shape with

Photo: H Burton

A basket of fruits of doum palm *Hyphaene thebaica* found in Tutankhamun's tomb.

rounded corners. When fresh, the smooth covering can be peeled off to expose the thin, very fibrous layer surrounding the large kernel. This layer is sweet and tastes of caramel – if gnawed with considerable patience! The fruits occur high up among the fan-shaped leaves which appear in tufts at the top of the 15 m (49 ft) tall slender trunks. These are always forked, unlike those of the date palm, which are unbranched. The leaves are of more importance than the fruits as they are widely used for basketry, ropes and similar fibrous items (see p 32).

Several species of doum palm with forked trunks have a wide distribution in the drier parts of tropical Africa south of the Sahara, but it is now considered that *H. thebaica* in the strict sense is limited to Sudan and Egypt. It was first described from Thebes in Egypt, hence the epithet *thebaica*. A few trees remain as far north as Elat at the end of the Gulf of Aqaba.

Juniperus species
Juniper
Family: Pinaceae   (see pl 38)

A large quantity of juniper berries occupied several baskets, either alone or mixed with dates, raisins, coriander and other seeds. When Boodle came to examine them he noticed two kinds, which have been identified as different species – both of them non-Egyptian. The prickly juniper (*J. oxycedrus*) is only a shrub or small tree, whereas the

Grecian juniper or eastern savin (*J. excelsa*) is a tree with a single straight trunk 20 m (65 ft) high and peeling bark.

The prickly juniper has three narrow, very sharply pointed (i.e. prickly) leaves about 10 mm (3/8 in) long, spreading out in all directions. Its fruit are smaller than the other species, being about 8 mm in diameter, dark brown in colour also with a yeasty bloom. They contain four to six seeds. Its distribution is from the mountains of south-east Europe to Central Asia and southwards only as far as Lebanon.

The more usually reported juniper species in Ancient Egypt is the Phoenician or brown-berried juniper (*J. phoenicea*) which occurs in Sinai, Jordan and around the Mediterranean sea, but its round red seeds were not found in Tutankhamun's tomb. Lucas indicates that berries of *J. drupacea* of Lebanon also occur in tombs.

Juniper berries are resinous and aromatic, hence their many uses since ancient times, especially medicinally and for mummification. They were incorporated between the layers of linen bandages around mummies and combined with natron, which preserved the flesh. Egyptian medical texts prescribe them as a diuretic and laxative. Their flavour was also imparted to foods such as stews; the pungent twigs flavoured grilled food. Juniper timber is durable, of good quality and has a wide range of uses. Moreover it is fragrant, and since it is reddish in colour, juniper wood may be that reported by Carter as the one used for various items of furniture in the tomb. Microscopic examination is required to confirm the identification of timbers and as far as I know none of the red wood has been studied, nor is it possible to do so at present. Lucas indicates that juniper wood (confusingly sometimes known as 'cedar' wood) was used for coffins and shrines. Juniper ('cedar') oil was employed for anointing dead bodies, so the presence of baskets of seeds in the tomb seems to have symbolic meaning.

Lens culinaris (synonym *L. esculenta*)
Lentil
Family: Leguminosae

Lentils dating from about 2000 BC have been excavated in Jericho and they are known to have been cultivated in Egypt by at least the time of the New Kingdom. Hence it is no surprise to find them included in Tutankhamun's tomb.

The slender annual vetch-like plants grow into one another, their tendrils forming a low tangle about 20 cm (8 in) high. The minute bluish pea flowers produce flat pods containing two lens-shaped seeds 3–5 mm (1/10–3/16 in) in diameter. Those in the tomb are large and were probably green lentils. Smaller seeds are usually reddish in colour, hence the red lentil stew ('mess of pottage') for which Esau exchanged his birthright (Genesis 25: 29–34). They are rich in protein and an important item of food in the dry countries in which lentils are cultivated.

A lentil plant *Lens culinaris*.

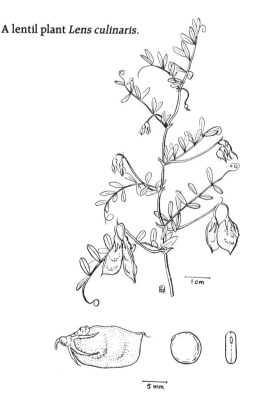

Black cumin *Nigella sativa*.

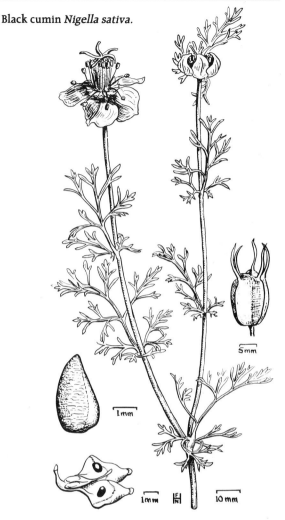

Nigella sativa
Black cumin
Family: Ranunculaceae

Seeds found in a pot proved to be those of the well-known black cumin which is related to the inedible love-in-the-mist (*N. damascena*) often grown as a blue-flowered annual in European gardens.

Black cumin (also known as nutmeg flower and gith) is an annual herb 30–40 cm (12–18 in) high with finely divided leaves. The terminal flowers are greenish, turning dull mauve, hence they are not very decorative. The erect horned capsules which appear later contain numerous black spicy seeds. These oily and aromatically piquant seeds are used for seasoning food and are sprinkled over bread.

To avoid spoiling the seeds during gathering they must be carefully threshed with sticks, not rolled over by a sledge like hard cereal grains (Isaiah 27: 27). The Ancient Assyrians and Copts found medicinal uses for the seeds, but it is not known whether the Egyptians also used them for treating ailments. The species is a native of north-west Africa and in the course of trade found its way to Egypt during the New Kingdom, since when it has been in regular cultivation, though seldom retrieved from excavations of tombs.

Phoenix dactylifera
Date palm
Family: Palmae   (see pl 39)

Many date fruits and stones were present in samples of fruits and seeds deposited in baskets in the tomb. The very sweet fruits of the date palm are still eagerly sought by the people of Egypt and perhaps the pharaoh himself was fond of them.

The date palm is only known in cultivation, although its wild relatives occur from the Canary Islands to eastern India. In Egypt it grows along the Nile and in desert oases such as Elim, where the Israelites found seventy trees (Exodus 15: 27). It is a tall tree growing to about 20 m (65 ft) high, with a slender unbranched trunk that is rough with the ridges left by the bases of old leaves. The leaves are enormous and feather-like, with a midrib and numerous lateral leaflets. Palm leaves were often depicted on the capitals of columns in temples. It was from the stout midrib that rafters and furniture were constructed, while the leaflets were used for thatching, basketry, sandals, nets and similar items (see p 32). The familiar date fruit, with sticky pulp surrounding a long hard stone, was eaten raw, often with milk, or fermented into wine. Date wine is referred to in ancient texts from the 2nd Dynasty onwards. Before the use of sugar was developed, dates and honey (see p 50) were popular sweeteners. Because they are so sweet, bacterial decay is inhibited, enabling cakes of dates to last for decades in an edible condition. For the same reason dates were useful as poultices and in some medicines.

Prunus dulcis (synonyms *P. amygdalus* or
Amygdalus communis)
Almond
Family: Rosaceae

About 30 almond stones, as well as some entire fruits, were found in two red pottery jars. The stones are very thick-walled and would have been difficult to crack in order to extract the fragrant 'nut'.

The almond tree thrives in the Mediterranean type of climate where they grow wild in maquis vegetation among small oak trees on rocky slopes facing south, such as occur in Palestine. It was from there that Jacob sent almond nuts as a gift to Joseph in Egypt where he was governor or vizier to the pharaoh of the day (Genesis 43: 11). Shells have often been excavated from ancient sites in Western Asia, but seldom in Egypt, where the tree was unlikely to be widely cultivated.

The very early flowering of the almond tree is well known, when the beautiful white or pink-tinged flowers cover the tree before its leaves appear. Its softly hairy fruits may be eaten whole when young and are seen for sale in Arab markets. However, they are usually left to mature, for the sake of the edible kernels. Some varieties have very bitter nuts, and these were used for the preparation of almond oil, which would have been used in cosmetics, and medicine. A small piece of its hard wood was used for one of Tutankhamun's arrows (see p 44).

Punica granatum
Pomegranate
Family: Lythraceae   (see pl 40)

A magnificent silver vase in the form of a pomegranate fruit, and another smaller one in ivory were included in the tomb contents.

A silver bowl in the form of a pomegranate fruit, found in Tutankhamun's tomb.

Photo: H Burton

Photo: A McRobb

Above: almond stones *Prunus dulcis* from Tutankhamun's tomb, now at Kew.

Below: a branch from the almond tree *Prunus dulcis.*

1 Cm

The pomegranate was no longer a novelty in Egypt at the time of Tutankhamun, since it had been introduced from the southern Caspian Sea region earlier in the 18th Dynasty, if not before. Its fresh leaves were included in a wreath 38 cm (15 in) long, together with willow leaves, and also in the elaborate floral collar with many other species (see p 9).

The pomegranate is an attractive plant which grows with several stems to form a rounded bush or small tree to about the height of a man. As a deciduous plant, it is leafless during the winter, and the rather narrow leathery leaves appear late in the spring. Flowering takes place in the hot season, when the scarlet flowers are large and brilliant. As the round fruits develop, the enlarged calyx persists on top. When ripe the hard rind is tinged pink, yellow and purple and it has long been used for dyeing leather. Inside the fruit are numerous seeds, each enclosed in white watery pulp refreshing to eat in the heat of summer.

In the tomb of Queen Hatshepsut's butler, Djehuty, dating from about 1470 BC – there is a large dry pomegranate together with gifts of flowers and other fruits. Pomegranates in bas-relief appear on the walls of Tuthmosis III's temple at Karnak dating from about 1450 BC, along with plants seen in Western Asia during his campaigns. It was in the same region that the Israelites used the form of the pomegranate to ornament the hem of the priestly robes (Exodus 28: 33–44) and later to decorate the capitals of the pillars of Solomon's temple (1 Kings 7: 20). As the pomegranate became more popular in Egypt its occurrence in tombs and as models increased in frequency.

Thymbra spicata
Wild thyme
Family: Labiatae

A large wooden box was found to contain a miscellaneous collection of objects – baskets, linen gloves, a palette, a throw-stick (boomerang) and also a sprig of wild thyme (*Thymbra spicata*) – evidently collected together during tidying up after robbers had entered the tomb.

This was an unexpected find since it is not an Egyptian plant, and one cannot imagine it being cultivated there. It occurs commonly in the eastern Mediterranean area as far south as the Judean mountains and rocky places on the coastal Plain of Sharon. Numerous branched stems about 30 cm (12 in) high, bear pairs of narrow leaves 1–2 cm (½–1 in) long. The small purplish flowers are grouped into clusters along the upper parts of the stems. The whole plant is fragrant owing to the presence of oil glands in the leaves.

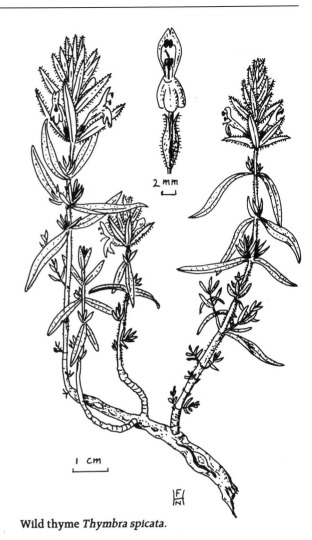

2 mm

1 cm

Wild thyme *Thymbra spicata.*

One presumes that it was placed there as a spicy herb with a pleasant smell and because of its rarity value in Egypt. No other record of it in tombs has been reported.

Trigonella foenum-graecum
Fenugreek
Family: Leguminosae

Fenugreek seeds were found mixed with coriander, lentils and other seeds in three baskets, a pot and even in a model granary. Tutankhamun's seeds are blue-black in colour, but when fresh, fenugreek seeds are yellow-brown. They sprout in two or three days and the shoots may be eaten as a nutritious salad, since they contain about 30 per cent protein as well as a fragrant, rather

Plate 36 A tree of sycomore fig *Ficus sycomorus* growing at Karnak.

Plate 37 (*bottom left*) Trees of doum palm *Hyphaene thebaica* with characteristic forked trunks, growing in the Yemen.

Plate 38 Branches of the prickly juniper *Juniperus oxycedrus*.

Photo: F N Hepper

Plate 39 A date palm *Phoenix dactylifera* in a wheat field in Kharga Oasis, Egypt.

Photo: F N Hepper

Plate 40 Spectacular flowers of the pomegranate bush *Punica granatum*.

bitter oil. The seeds are borne in narrow, sickle-like beaked pods up to 12 cm (4½ in) long. The parent plant is an annual legume, with few branches 20 to 50 cm (8–20 in) high, bearing trifoliate leaves which are saw-toothed. One or two small, whitish pea flowers occur in the axils of the leaves in March and April.

Fenugreek is indigenous in the eastern Mediterranean region and was known to the Ancient Egyptians in pre-dynastic times, about 3000 BC. It is still widely cultivated in dry countries such as the Yemen and India, where it is grown for both fodder and for human consumption as a pot-herb. Some leaves I collected and dried in the Yemen years ago are still aromatic. Fenugreek seeds, besides being eaten cooked with other foods, are also useful medicinally as an ointment and to encourage lactation.

Below: **Fenugreek** *Trigonella foenum-graecum*.

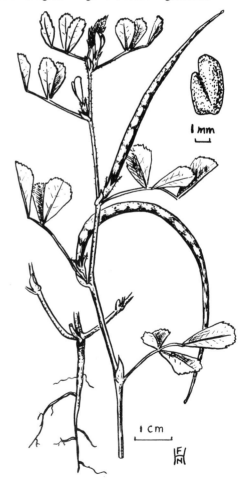

Photo: H Burton

The shoot of wild thyme *Thymbra spicata* found in Tutankhamun's tomb.

Triticum dicoccum
Emmer wheat
Family: Gramineae

Wheat, like barley, is an annual grass. There are several
kinds of wheat, and emmer is one that is not now grown,
except perhaps in remote parts of Ethiopia and Italy. In
ancient times up to about the Roman period it was the
principal kind of wheat. In the Mediterranean region,
hard wheat (*T. durum*), which is similar to emmer, is still
used for a pitta-style bread and for pasta, but nowadays
the so-called bread wheat (*T. aestivum*) is the usual crop
for flour world-wide.

As a grass it germinates with one seed leaf (mono-
cotyledon) and sends up one to four stems bearing several
very narrow leaves. During the late spring in Egypt a
green head develops on each stem. These are the 'ears',
which are whiskery with awns, like barley but unlike
most forms of bread wheat nowadays. As the grains swell
and ripen, the leaves and stems turn yellow and become
straw, which the Israelites had to find for their bricks
when they were oppressed by pharaoh (Exodus 5:7).

**Grains of emmer wheat *Triticum dicoccum* from
Tutankhamun's tomb, now at Kew.**

Photo: A McRobb

1 cm

Above: An ear of bread wheat *Triticum aestivum*.

Several grains are grouped into small clusters (spikelets) which break off from the ear on threshing (this is carried out with a flail or a sledge drawn by animals), and then are separated from one another and their chaff removed by further threshing. A symbolic flail was held by Tutankhamun's golden effigy on his mummiform coffin, as well as a crook; these were typical of Osiris as cultivator (flail) and shepherd (crook). Winnowing removed the edible kernel from the light scaly chaff.

During the time of Joseph in Egypt, Genesis 41 relates that the pharaoh, who has not been positively identified, had a dream in which he saw 'seven ears growing on one stalk, full and good' signifying the seven years of plenty. Although this is an allegory, Vivi Täckholm noted that there is still a seven-headed wheat, recognised by the Arabs from the branching of its ear. This is a variety of the rivet wheat (*T. turgidum*), which is very similar to the hard wheat mentioned above. The identity of the cereal which was not ruined by hail, mentioned in Exodus 9: 32, is also a problem – it cannot be rye (*Secale cereale*), the 'rie' of the Authorized Version (the King James Bible), nor 'spelt' (*T. spelta*) of other versions, but it probably was also a variety of emmer akin to the hard wheat.

Vitis vinifera
Grape vine
Family: Vitaceae

Several baskets contained various fruits, including shrivelled grapes and their seeds. Grapes would have been eaten immediately after harvesting and the large pips are often found in excavations of old buildings, as they are quite resistent to decay. Surplus grapes would have been sun-dried to keep as raisins. Analysis of fresh raisins shows that they are rich in sugar, which can amount to as much as 70 per cent of their weight. They keep for quite a long time in an edible condition but eventually they shrivel into hard grains, so it is difficult to decide whether those found in Tutankhamun's tomb were placed there as dry raisins or as fresh grapes.

The vine is a climbing shrub with no distinct trunk. The stem thickens with age and as the long growths are severely cut back each year, the perennial stem is readily distinguishable by its rough brown bark from the trailing annual shoots. These grow with remarkable speed during the summer months, having started from a bud on the old stem. As the bud swells it produces a shoot bearing five to

Grapes *Vitis vinifera*.

Photo: F N Hepper

seven divided (palmate) leaves which develop on alternate sides of the shoot. Opposite the third, fourth and sometimes the fifth leaf a cluster of flowers is produced, beyond the fifth or sixth leaf only branched tendrils develop opposite them. Each tiny flower has green petals which hold together as a little cap and fall in one piece when the air reaches a certain temperature. The stamens spread out, shedding pollen on to the pistils of its own and neighbouring flowers to effect pollination. Wild ancestral vines, however, have male and female flowers on separate plants.

After the fertilisation of the flowers, fruits develop as small green spheres, which are bitter until they ripen in late summer. They are then either green or red, and very sweet and full of juice, which is dried out to make them into raisins or pressed out to make wine (pp 50–1).

Ziziphus spina-christi
Christ-thorn
Family: Rhamnaceae

Numerous whole Christ-thorn fruits and seeds – the latter probably from decayed fruits – occurred in various baskets and pots, usually mixed with lentils, dates and other fruits, as well as in the loaves (p 53). Carter used the Arabic name *nabuk* for these fruits: the tree is called *sidder*.

Fresh Christ-thorn fruits are about the size of an olive, but yellow and containing two or three hard seeds. The outer layer is edible and sweet or astringent when ripe. Immature fruits are said to be medicinally useful as a laxative and febrifuge. They grow on a spiny, evergreen bush or small tree like a hawthorn with a number of stems or side branches, and twigs which are rather zigzag-shaped and whitish in appearance.

Its timber is sometimes used, as it is hard and durable, hence its occurrence as cross-tongues in Tutankhamun's shrines (see p 38) and for plywood coffins dating from the 3rd Dynasty. The alternate, stalked, oval leaves have small teeth along the margin and the blade is conspicuously three-nerved from the base. A pair of very sharp unequal thorns grow at the base of each leaf-stalk (except in some cultivated trees), hence it was said to have been used for Jesus' crown of thorns (John 19: 2). The flowers are pale yellow, small and inconspicuous.

Christ-thorn was one of the most important and well-known plants of Ancient Egypt, being native to the country and useful in so many ways, as food, medicine and timber. It also had religious significance.

*   *   *

For other fruits, see: mandrake (*Mandragora officinarum* (p 15) and persea (*Mimusops laurifolia* p 15).

Flowering and fruiting shoots of Christ-thorn *Ziziphus spina-christi*.

Further Reading

Plant Species

Abies cilicica Ant. & Kotschy (Cilician fir, p 44)
Hepper, F N *Planting a Bible Garden*, London: HMSO, 1987, p 63, fig 46a; Lucas, A *Ancient Egyptian Materials and Industries*, London: Arnold, 1926 (J R Harris 4th ed. 1962), pp 319, 320, 436; Meiggs, R *Trees and Timber in the Ancient Mediterranean World* Oxford: Clarendon Press, 1982, fig 15b (shown as *A. alba*).

Acacia species (acacia, p 22)
Germer, R *Flora des Pharaonischen Ägypten*, Mainz am Rhein: Philipp von Zabern, 1985, pp 89–92; Hepper *Planting a Bible Garden*, p 78, fig 60, pl 33; Lucas *Materials*, pp 5, 34, 47, 440–41; Täckholm, V *Students' Flora of Egypt*, Cairo: Cairo University, 1974, pp 289–91, pl 93; Winlock, H E 'Materials used at the embalming of King Tut-Ankh-Amun' in *Metropolitan Museum of Art Papers* No 10, 1941, pp 5–18; Woenig, F *Die Pflanzen in Alten Aegypten*, Leipzig: Friedrich, 1886, pp 298–304; Zohary, M *Flora Palaestina*, Jerusalem: Israel Academy of Sciences, 1972, vol 2, pp 26–8, pl 38–41.

Allium sativum L. (garlic, p 55)
Darby, W, Ghalioungui, P and Grivetti, L *Food: the Gift of Osiris*, London: Academic Press, 1977, vol 2, pp 650–60; Germer *Flora*, pp 194–95; Germer, R *Die Pflanzenmaterialien aus dem Grab des Tutanchamun*, Hildersheim: Gerstenberg, 1989, pp 43–4, pl 8; Laurent-Täckholm, V *Faraos blomster*, Stockholm: Natur och Kultur, 1951, pp 159–61; Manniche, L *An Ancient Egyptian Herbal*, London: British Museum, 1989, pp 70–71; Zohary, D and Hopf, M *The Domestication of Plants in the Old World*, Oxford: Clarendon Press, 1988, p 169.

Anthemis pseudocotula Boiss. (mayweed, camomile, p 13)
Feinbrun-Dothan, N *Flora Palaestina*, Jerusalem: Israel Academy of Sciences, 1978, vol 3, p 338, pl 52; Täckholm *Students' Flora*, p 574, pl 204A.

Apium graveolens L. (wild celery, p 14)
Germer *Flora*, pp 137–8; Germer *Pflanzenmaterialien*, p 8; Laurent-Täckholm *Faraos blomster*, pp 110–11; Täckholm *Students' Flora*, p 388; Zohary *Flora Palaestina*, vol 2, p 416, pl 600.

Balanites aegyptiaca L. (Egyptian plum, p 23)
Germer *Flora*, pp 98–100; Germer *Pflanzenmaterialien*, pp 37–45; Laurent-Täckholm *Faraos blomster*, pp 232–3; Lucas *Materials*, pp 331, 499; Manniche *Herbal*, p 81; Woenig *Pflanzen*, pp 319–21.

Betula pendula Roth (silver birch, p 45)
Davis, P H *Flora of Turkey*, Edinburgh: University Press, 1982, vol 7, p 690; Germer
Flora, p 18; Germer *Pflanzenmaterialien*, p 78; Lucas *Materials*, pp 454–5.

Boswellia sacra Flueck. (frankincense, p 23)
Germer *Flora*, pp 108–11; Hepper *Planting a Bible Garden*, p 80, fig 62; Hepper, F N
'Arabian and African frankincense trees', *Journal of Egyptian Archaeology*, 1969, vol 55,
pp 66–72; Lucas *Materials*, pp 91–2.

Carthamus tinctorius L. (safflower, p 32)
Germer *Flora*, pp 173–5; Germer *Pflanzenmaterialien*, p 37; Laurent-Täckholm *Faraos
blomster*, pp 232–5; Lucas *Materials*, p 153; Manniche *Herbal*, p 83; Zohary and Hopf
Domestication, p 174.

Cedrus libani A Rich. (cedar of Lebanon, p 45)
Germer *Flora*, pp 5–6; Germer *Pflanzenmaterialien*, pp 71–5; Hepper, *Planting a Bible
Garden*, p 63, fig 47, pl 27; Laurent-Täckholm *Faraos blomster*, pp 236–9; Meiggs *Trees
and Timber*, pp 49–87.

Centaurea depressa Bieb. (cornflower, p 14)
Germer *Flora*, p 173; Germer *Pflanzenmaterialien*, pp 5–19; Manniche *Herbal*, p 85.

Cicer arietinum L. (chick-pea, p 55)
Darby *Food*, vol 2, pp 685–7, fig 17.11; Germer *Flora*, pp 96–7; Germer
Pflanzenmaterialien, p 36; Zohary and Hopf *Domestication*, pp 98–102.

Citrullus lanatus (Thunb.) Mansf. (water-melon, p 56)
Darby *Food*, vol 2, pp 717–18, fig 18.8; Germer *Flora*, pp 127–8; Germer
Pflanzenmaterialien, p 56; Hepper *Planting a Bible Garden*, p 17, fig 6; Manniche *Herbal*,
p 92; Woenig *Pflanzen*, pp 200–207; Zohary and Hopf *Domestication*, pp 167–168.

Cocculus hirsutus (J R & G Forst.) Diels (cocculus, p 56)
Germer *Flora*, pp 36–37; Germer *Pflanzenmaterialien*, pp 54–5. *C. pendulus*: Täckholm
Students' Flora, p 144, pl 38; Zohary *Flora Palaestina*, vol 1, p 216, pl 317.

Commiphora gileadensis (L.) Chr. (balm of Gilead, p 24)
Germer *Flora*, pp 107–8; Hepper *Planting a Bible Garden*, p 84, fig 66a.

Commiphora myrrha (Nees) Engl. (myrrh, p 24)
Germer *Flora*, pp 106–7; Hepper *Planting a Bible Garden*, p 84, fig 66; Laurent-
Täckholm *Faraos blomster*, pp 175–85; Lucas *Materials* pp 92–3.

Coriandrum sativum L. (coriander, p 57)
Darby *Food*, vol 2, pp 798–9; Germer *Flora*, pp 135–6; Germer *Pflanzenmaterialien*, p 60;
Hepper *Planting a Bible Garden*, p 16, fig 5, pl 4; Manniche *Herbal*, p 94; Täckholm
Students' Flora, p 385; Woenig *Pflanzen*, p 225; Zohary and Hopf *Domestication*, p 172.

Cupressus sempervirens L. (Cypress, p 46)
Germer *Flora* pp 9–10; Germer *Pflanzenmaterialien*, p 76; Hepper *Planting a Bible
Garden*, p 64, fig 48, pl 28; Täckholm, V and Drar, M *Flora of Egypt*, Cairo: 1941, Cairo
University Press, vol 1, p 64; Zohary *Flora Palaestina*, vol 1, p 19, pl 17.

Cyperus papyrus L. (papyrus sedge, p 33)
Germer *Flora*, pp 248–50; Germer *Pflanzenmaterialien*, p 65; Hepper *Planting a Bible
Garden*, p 73, fig 56; Hepper, F N and Reynolds, T 'Papyrus and the adhesive properties
of its cell sap in relation to paper-making', *Journal of Egyptian Archaeology*, 1967,
vol 53, pp 156–7; Laurent-Täckholm *Faraos blomster*, p 15–42; Lucas *Materials*,

pp 137–40; Manniche *Herbal*, pp 99–100; Täckholm and Drar *Flora of Egypt*, vol 1, pp 15–18; Täckholm *Students' Flora*, p 790; Woenig *Pflanzen*, pp 74–135, fig 62.

Dalbergia melanoxylon Guill. & Perr. (ebony, p 46)
Germer *Flora*, pp 97–8; Germer *Pflanzenmaterialien*, pp 71–2; Hepper, F N 'On the transference of ancient plant names' in *Palestine Exploration Quarterly*, 1977, vol 109, pp 129–30; Lucas *Materials*, pp 434–6.

Desmostachya bipinnata (L.) Stapf (halfa grass, p 33)
Feinbrun-Dothan *Flora Palaestina*, vol 4, p 289, pl 383; Germer *Flora*, p 202; Germer *Pflanzenmaterialien*, p 68; Täckholm and Drar *Flora of Egypt*, vol 1, pp 177–85; Täckholm *Students' Flora*, p 690, pl 255A.

Ficus sycomorus L. (sycomore fig, p 58)
Darby *Food*, vol 2, pp 745–8, fig 18.25; Galil, J 'An ancient technique for ripening sycomore fruits in East-Mediterranean countries' in *Economic Botany*, New York, 1968, vol 22, pp 178–90; Germer *Flora*, pp 25–7; Germer *Pflanzenmaterialien* pp 55–6; Hepper *Planting a Bible Garden*, p 47, fig 33; Laurent-Täckholm *Faraos blomster*, pp 43–51; Lucas *Materials*, p 447; Manniche *Herbal*, pp 103–5; Zohary and Hopf *Domestication*, pp 145–6.

Fraxinus species (ash, p 47)
Feinbrun-Dothan *Flora Palaestina*, vol 3, p 16, pl 21; Germer *Flora*, p 152; Germer *Pflanzenmaterialien*, p 76.

Grewia tenax (Forssk.) Fiori (grewia, p 59)
Germer *Flora*, p 19; Germer *Pflanzenmaterialien*, p 54; Täckholm *Students' Flora*, p 348, pl 118.

Hordeum vulgare L. (barley, p 59)
Darby *Food*, vol 2, pp 479–89; Germer *Flora*, pp 207–10; Germer *Pflanzenmaterialien*, pp 26–33; Hepper *Planting a Bible Garden*, p 18, fig 7, pl 6; Manniche *Herbal*, pp 107–8; Woenig *Pflanzen*, pp 136–50; Zohary and Hopf *Domestication*, pp 52–63.

Hyphaene thebaica (L.) Mart. (doum palm, p 59)
Darby *Food*, vol 2, pp 730–33, figs 18.16, 18.17; Germer *Flora*, pp 234–5; Germer *Pflanzenmaterialien*, pp 47–8, pl 9; Laurent-Täckholm *Faraos blomster*, pp 207–12; Täckholm and Drar *Flora of Egypt*, vol 2, p 273ff; Täckholm *Students' Flora*, p 763; Woenig *Pflanzen*, pp 317–18.

Imperata cylindrica (L.) Beauv. (imperata halfa grass, p 33)
Feinbrun-Dothan *Flora Palaestina*, vol 4, p 318, pl 423; Germer *Flora*, pp 224–5; Lucas *Materials*, p 129; Täckholm and Drar *Flora of Egypt*, vol 1, pp 482–6; Täckholm *Students' Flora*, p 757, pl 279.

Juncus arabicus (Asch. & Buch.) Adamson (rush, p 34)
Feinbrun-Dothan *Flora Palaestina*, vol 4, p 141, pl 187; Germer *Flora*, p 200; Germer *Pflanzenmaterialien*, pp 68–9; Täckholm and Drar *Flora of Egypt*, pp 462–4; Täckholm *Students' Flora*, p 664.

Juniperus species (juniper, p 60)
Darby *Food*, vol 2, p 716; Germer *Flora*, pp 10–12; Germer *Pflanzenmaterialien*, pp 57–60; Lucas *Materials*, pp 310–12, 437; Manniche *Herbal*, pp 110–12; Zohary *Flora Palaestina*, vol 1, p 20, pl 18.

Lawsonia inermis L. (henna, p 25)
Germer *Flora*, p 126; Hepper, *Planting a Bible Garden*, p 86, fig 69; Lucas *Materials*, p 310; Manniche *Herbal*, p 114.

Lens culinaris Medik. (lentil, p 61)
Darby *Food*, vol 2, pp 687–9; Germer *Flora*, pp 86–7; Germer *Pflanzenmaterialien*,
pp 34–5; Manniche *Herbal*, p 115; Woenig *Pflanzen*, p 214; Zohary and Hopf
Domestication, pp 85–92.

Lilium candidum L. (white lily, p 25)
Feinbrun-Dothan *Flora Palaestina*, vol 4, p 44, pl 63; Germer *Flora*, pp 195–6; Hepper
Planting a Bible Garden, p 32, fig 20; Laurent-Täckholm *Faraos blomster*, pp 168–9.

Linum usitatissimum L. (flax, p 34)
Darby *Food*, vol 2, pp 783–4; Germer *Flora*, pp 100–101; Germer *Pflanzenmaterialien*,
p 37; Hepper *Planting a Bible Garden*, p 19, fig 8, pl 7; Laurent-Täckholm *Faraos
blomster*, pp 252–69; Lucas *Materials*, pp 142–6; Manniche *Herbal*, p 116; Pfister, R 'Les
textiles du tombeau de Toutankhamon' *Revue des Arts Asiatiques*, 1937, vol 11,
pp 211–15; Täckholm and Drar *Flora of Egypt*, vol 2, pp 252–69; Woenig *Pflanzen*,
pp 181–9; Zohary and Hopf *Domestication*, pp 114–19.

Liquidambar orientalis Mill. (Levant storax tree, p 47)
Germer *Flora*, p 65; Germer *Pflanzenmaterialien*, p 77; Hepper *Planting a Bible Garden*,
p 49, fig 35; Lucas *Materials*, pp 95, 437.

Mandragora officinarum L. (mandrake, p 15)
Feinbrun-Dothan *Flora Palaestina*, vol 4, p 167, pl 278; Germer *Flora*, pp 169–71;
Germer *Pflanzenmaterialien* pp 11–12; Hepper *Planting a Bible Garden*, p 34, fig 22;
Laurent-Täckholm *Faraos blomster*, pp 96–9.

Mimusops laurifolia (Forssk.) Friis (persea, p 15)
Darby *Food*, vol 2, pp 736–40; Germer *Flora*, p 148; Germer *Pflanzenmaterialien*,
pp 11–18, pl 5, 6; Friis, I, Hepper, F N and Gasson, P 'The botanical identity of the
Mimusops in Ancient Egyptian tombs' *Journal of Egyptian Archaeology*, 1986, vol 72,
pp 201–2; Laurent-Täckholm *Faraos blomster*, pp 120–23; Lucas *Materials*, p 445;
Manniche *Herbal*, pp 121–2.

Moringa peregrina (Forssk.) Fiori (horseradish tree, p 25)
Darby *Food*, vol 2, p 784; Germer *Flora*, pp 58–9; Lucas *Materials*, pp 331–2; Täckholm
Students' Flora, p 211, pl 65; Zohary *Flora Palaestina*, vol 1, p 340, pl 495.

Nigella sativa L. (black cumin, p 61)
Darby *Food*, vol 2, p 807; Germer *Flora*, p 35; Germer *Pflanzenmaterialien*, p 62; Hepper
Planting a Bible Garden, p 21, fig 10; Manniche *Herbal*, p 125; Täckholm *Students' Flora*,
p 385; Zohary and Hopf *Domestication*, p 173.

Nymphaea caerulea Savigny (blue lotus waterlilly, p 16)
Germer *Flora*, pp 37–9; Laurent-Täckholm *Faraos blomster*, pp 127–34; Täckholm
Students' Flora, p 144, pl 39.

Nymphaea lotus L. (white lotus waterlily, p 16)
In addition to references for *Nymphaea caerulea* (above): Hepper *Planting a Bible
Garden*, p 74, fig 57; Manniche *Herbal*, pp 126–9; Zohary *Flora Palaestina*, vol 1, p 217,
pl 319.

Olea europaea L. (olive, p 16)
Darby *Food*, vol 2, pp 718–21; Germer *Flora*, pp 150–51; Germer *Pflanzenmaterialien*,
pp 5–8, 37; Hepper *Planting a Bible Garden*, p 54, fig 40, pl 23; Lucas *Materials*,
pp 333–5; Woenig *Pflanzen*, pp 227; Zohary and Hopf *Domestication*, pp 131–6.

Papaver rhoeas L. (corn poppy, p 16)
Germer *Flora*, p 44; Hepper *Planting a Bible Garden*, p 22, fig 11; Manniche *Herbal*, p 130; Täckholm *Students' Flora*, p 153.

Phoenix dactylifera L. (date palm, p 62)
Darby *Food*, vol 2, pp 722–30, pl 18.12–15; Germer *Flora*, pp 232–3; Germer *Pflanzenmaterialien*, p 45; Hepper *Planting a Bible Garden*, p 87, pl 38, 39; Laurent-Täckholm *Faraos blomster*, pp 194–201; Lucas *Materials*, p 443; Manniche *Herbal*, pp 133–4; Täckholm and Drar *Flora of Egypt*, vol 2, pp 218 ff; Täckholm *Students' Flora*, p 763; Woenig *Pflanzen*, pp 304–14; Zohary and Hopf *Domestication*, pp 146–50.

Phragmites australis (Cav.) Trin. ex Steud. (common reed, p 35)
Feinbrun-Dothan *Flora Palaestina*, vol 4, p 270, pl 353; Germer *Flora*, pp 203–206; Germer *Pflanzenmaterialien*, pp 20–21; Hepper *Planting a Bible Garden*, p 75, fig 58; Täckholm and Drar *Flora of Egypt*, vol 1, p 213; Täckholm *Students' Flora*, p 696.

Picris radicata (Forssk.) Less. (ox-tongue, p 16)
Feinbrun-Dothan *Flora Palaestina*, vol 3, pp 419, pl 707; Germer *Flora*, p 184; Germer *Pflanzenmaterialien*, pp 11, 25; Täckholm *Students' Flora*, p 597, pl 216.

Pinus species (pine, p 26)
Darby *Food*, vol 2, p 784; Germer *Flora*, p 8; Germer *Pflanzenmaterialien*, p 72; Hepper *Planting a Bible Garden*, p 67; fig 51, pl 29; Lucas *Materials*, p 319.

Pistacia species (mastic and Chios balm, p 26)
Germer *Flora*, pp 111–13; Hepper *Planting a Bible Garden*, p 88, fig 71; Mills, J S and White, R 'The identity of the resins from the late Bronze Age shipwreck at Ulu Burun (Kas)', *Archaeometry*, 1989, vol 31, pp 37–44; Stol, M *On trees, mountains and millstones in the Ancient Near East*, Leiden: Ex Oriente Lux, 1979, pp 1–24.

Prunus dulcis (Mill.) D A Webb (almond, p 62)
Darby *Food*, vol 2, pp 780–82; Germer *Flora*, p 58; Germer *Pflanzenmaterialien*, p 38; Hepper *Planting a Bible Garden*, p 55, fig 41; Lucas *Materials*, p 329; Manniche *Herbal*, pp 138–9; Zohary and Hopf *Domestication*, pp 160–61.

Punica granatum L. (pomegranate, p 62)
Darby *Food*, vol 2, pp 740–44, fig 18.23–4; Germer *Flora*, pp 42–3; Germer *Pflanzenmaterialien*, pp 11–15; Hepper *Planting a Bible Garden*, p 56, fig 42; Laurent-Täckholm *Faraos blomster*, pp 220–23; Lucas *Materials*, pp 35–6; Woenig *Pflanzen*, pp 323–30; Zohary and Hopf *Domestication*, pp 150–51.

Quercus aegilops L. (Valonia oak, p 48)
Germer *Flora*, pp 20–21; Germer *Pflanzenmaterialien*, p 74; Hepper *Planting a Bible Garden*, p 69; Lucas *Materials*, p 438; Meiggs *Trees and Timber*, p 45; Zohary *Flora Palaestina*, vol 1, p 32, pl 31.

Ricinus communis L. (castor oil plant, p 26)
Darby *Food*, vol 2, pp 782–3; Germer *Flora*, pp 103–4; Germer *Pflanzenmaterialien*, p 42; Hepper *Planting a Bible Garden*, p 89, pl 40; Lucas *Materials*, p 332; Manniche *Herbal*, pp 142–3.

Rubia tinctorum L. (madder, p 35)
Feinbrun-Dothan *Flora palaestina*, vol 3, p 235, pl 395; Lucas *Materials*, p 153; Manniche *Herbal*, p 144; Pfister, R 'Les Textiles', p 209.

Salix subserrata Willd. (willow, p 17)
Germer *Flora*, pp 16–17; Germer *Pflanzenmaterialien*, pp 8–18; Täckholm *Students' Flora*, p 51, pl 58.

Sesamum indicum L. (sesame, p 27)
Darby *Food*, vol 2, pp 497–8; Germer *Flora*, pp 171–2; Germer *Pflanzenmaterialien*, p 41; Lucas *Materials*, p 336; Manniche *Herbal*, p 147; Täckholm *Students' Flora*, p 503, pl 176; Zohary and Hopf *Domestication*, pp 126–7.

Tamarix aphylla (L.) Karst. (tamarisk, p 48)
Germer *Flora*, pp 124–5; Germer *Pflanzenmaterialien*, p 77; Manniche *Herbal*, pp 149–50; Täckholm *Students' Flora*, p 366, pl 125; Woenig *Pflanzen*, pp 341–3; Zohary *Flora Palaestina*, vol 2, p 359, pl 522.

Thymbra spicata L. (wild thyme, p 64)
Feinbrun-Dothan *Flora Palaestina*, vol 3, p 152, pl 252; Germer *Pflanzenmaterialien*, p 62, pl 10.

Trigonella foenum-graecum L. (fenugreek, p 64)
Darby *Food*, vol 2, pp 801–2; Germer *Flora*, p 68; Germer *Pflanzenmaterialien*, p 151; Zohary *Flora Palaestina*, vol 2, p 135, pl 199; Zohary and Hopf *Domestication*, pp 110–11.

Triticum dicoccum Schrank (emmer wheat, p 66)
Darby *Food*, vol 2, pp 486–92; Germer *Flora*, pp 210–11; Germer *Pflanzenmaterialien*, pp 26–33; Laurent-Täckholm *Faraos blomster*, pp 57–61; Manniche *Herbal*, pp 152–3; Woenig *Pflanzen*, pp 136–80; Zohary and Hopf *Domestication*, pp 16–52.

Typha domingensis Pers. (reed-mace, p 36)
Feinbrun-Dothan *Flora Palaestina*, vol 4, p 345, pl 451; Germer *Flora*, p 241; Hepper *Planting a Bible Garden*, p 76, fig 59; Täckholm and Drar *Flora of Egypt*, vol 1, p 90; Täckholm *Students' Flora*, p 770, pl 284.

Ulmus minor Mill. (elm, p 49)
Germer *Flora*, p 22; Germer *Pflanzenmaterialien*, p 77; Lucas *Materials*, p 436.

Vitis vinifera L. (grape vine, p 67)
Černý, J, *Tutankhamun Tomb Series*, Oxford: Griffith Institute 1965, vol 2, pp 1–7; Darby *Food*, vol 2, pp 551–618; Germer *Flora*, pp 116–18; Germer *Pflanzenmaterialien*, pp 49–50; Hepper *Planting a Bible Garden*, pp 60–61, fig 46, pl 28; Laurent-Täckholm *Faraos blomster*, pp 224–9; Lesko, L H *King Tut's wine cellar*, Berkeley: University of California 1977; Lucas *Materials*, pp 16–22; Manniche *Herbal*, pp 155–6; Woenig *Pflanzen*, pp 254–76; Zohary and Hopf *Domestication*, pp 136–42.

Withania somnifera (L.) Dunal (withania nightshade, p 18)
Feinbrun-Dothan *Flora Palaestina*, vol 3, p 164, pl 272; Germer *Flora*, p 167; Germer *Pflanzenmaterialien*, pp 12, 25; Täckholm *Students' Flora*, p 474, pl 164.

Ziziphus spina-christi (L.) Desf. (Christ-thorn, p 68)
Germer *Flora*, p 114; Germer *Pflanzenmaterialien*, pp 50–51; Hepper *Planting a Bible Garden*, p 87, fig 73; Lucas *Materials*, p 446; Täckholm *Students' Flora*, p 345, pl 117c; Zohary *Flora Palaestina*, vol 2, p 307, pl 450; Zohary and Hopf *Domestication*, p 178.

Books on Tutankhamun

Carter, H and Mace, A C *The Tomb of Tut-Ankh-Amen*, London: Cassell 1923–33, vols 1–3. (Newberry, P E Appendix III, Report on the floral wreaths found in the coffins of Tut-Ankh-Amen, vol 2, pp 189–96). See also: *Tutankhamun Tomb Series*, Oxford: Griffith Institute, 1963–, vols 1–8.

Desroches-Noblecourt, C *Tutankhamen*, London: *The Connoisseur* and Michael Joseph, 1963.

Edwards, I E S *Tutankhamen, his tomb and its treasures*, London: Penguin, 1979.

Germer, R *Die Pflanzenmaterialien aus dem Grab des Tutanchamun*, Hildersheim: Gerstenberg, 1989.

Neubert, O *Tutankhamun and the Valley of the Kings*, London: Mayflower, 1972.

Reeves, N *The Complete Tutankhamun*, London: Thames & Hudson, 1990.

Index

Page numbers in *italics* refer to black and white illustrations and those in **bold** refer to main descriptions of plants.
Colour plates are identified as 'pl'.